Fundamentals of Basic Mathematical Tools

(Class I-VIII)

Fundamentals of Basic Mathematical Tools

(Class I-VIII)

G.N. Tiwari and Neha Dimri

Notion Press

Old No. 38, New No. 6
McNichols Road, Chetpet
Chennai - 600 031

First Published by Notion Press 2016
Copyright © G.N. Tiwari and Neha Dimri 2016
All Rights Reserved.

ISBN 978-1-945579-38-7

To, our respected guru ji,

Padma Shri Prof. Mahendra Singh Sodha, FNA

Prof. G.N. Tiwari (left) discussing with
Prof. Mahendra Singh Sodha (right)

Contents

Preface

We felt the need to write this booklet for class I-VIII students to facilitate the understanding of basic mathematics from the beginning itself. After visiting many places and interacting with numerous primary students, we found that there is a problem not only with the students but teachers as well. We are sure that it is only due to lack of basic understanding of mathematics, that the students are running away from the study of science and engineering. In order to address this problem we have made an attempt to place the basic mathematical tools in a simpler form for class I-VIII to encourage the younger kids, so that they could understand that mathematics is not difficult. All it needs is courage to start. This booklet is based on our learning experience from primary school to higher education level.

We hope that the teachers as well as parents will also acquire benefits from this booklet while encouraging the students to adopt mathematics from the initial stage. We have also included exercises at the end of the chapters, in this booklet to help the students test the knowledge grasped. Further, we request the users of this booklet to suggest any improvements on this booklet for future course of action.

We thank all the primary teachers/students, including Shri Vats Tiwari, Class V, Presidium School, Indirapuram, Ghaziabad (UP), who gave their inputs in improving the content of this booklet. We have also utilized various online resources for the development of this booklet.

– G.N. Tiwari and Neha Dimri

Bag Energy Research Society (www.bers.in)
Prodyogiki Apartment,
House No. A-112, Sector-3, Dwarka,
New Delhi-110075

11 B, Gyan Khand 4,
Gaur Buildings, Indirapuram, Ghaziabad,
Uttar Pradesh-201010

Alphabets

1.1 Hindi Alphabets

स्वर

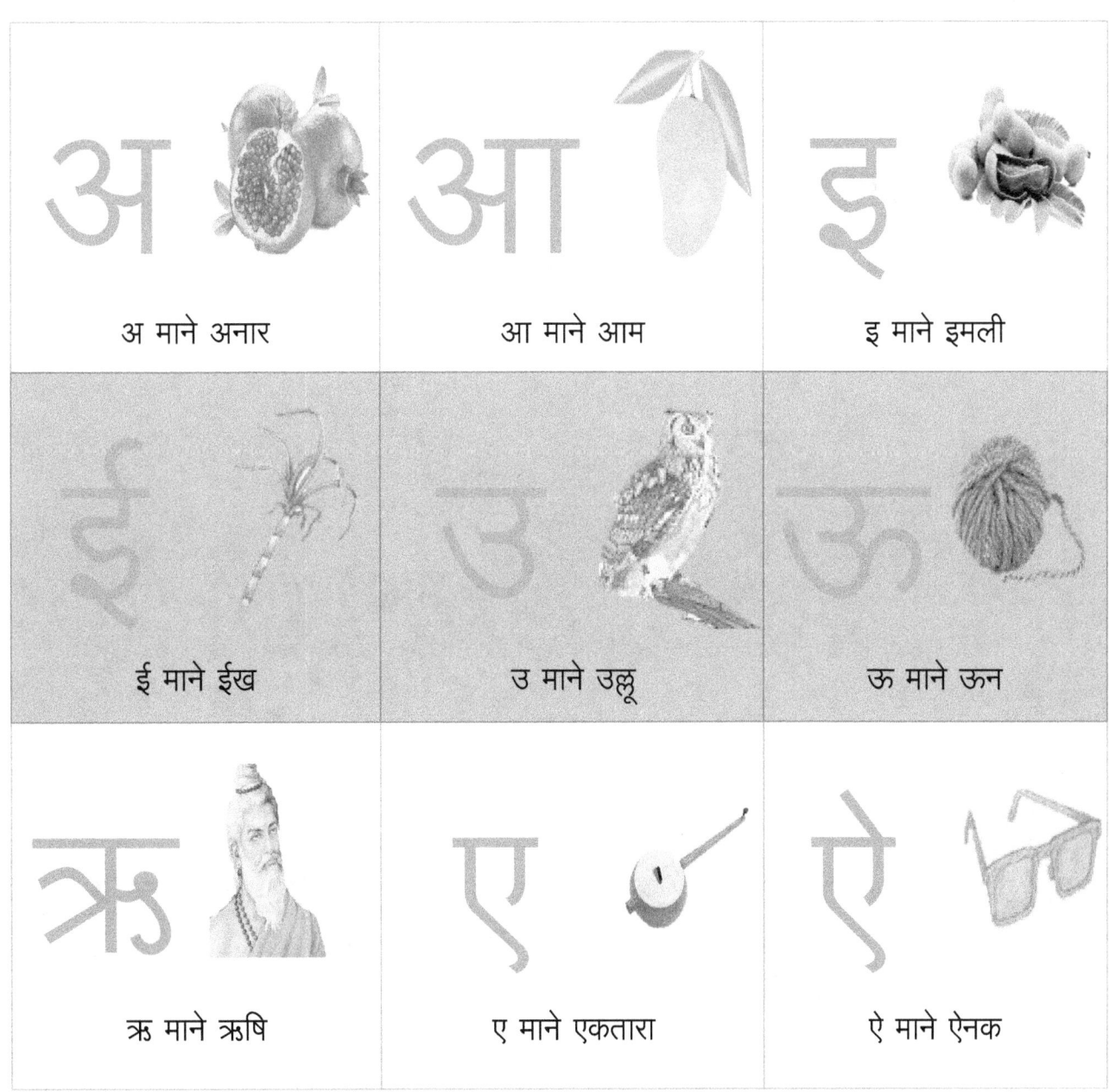

अ माने अनार

आ माने आम

इ माने इमली

ई माने ईख

उ माने उल्लू

ऊ माने ऊन

ऋ माने ऋषि

ए माने एकतारा

ऐ माने ऐनक

ओ माने ओखली   औ माने औरत   अं माने अंगूर

अः

अः माने कुछना

व्यंजन

क माने कबूतर

ख माने खरगोश

ग माने गमला

घ माने घर

ङ माने कुछना

च माने चम्मच

छ माने छतरी

ज माने जहाज

झ माने झण्डा

ञ माने कुछना

ट माने टमाटर

ठ माने ठठेरा

ड माने डमरू

ढ माने ढक्कन

ण माने कुछना

त माने तरबूज

थ माने थन

द माने दवात

ध माने धनुष

न माने नल

प माने पतंग

फ माने फल

ब माने बन्दर

भ माने भालू

म माने मछली

य माने यज्ञ

र माने रथ

ल माने लट्टू

व माने वनमानुष

श माने शरीफा

ष माने षट्कोण

स माने सपेरा

ह माने हवाई जहाज

क्ष माने क्षत्रिय   त्र माने त्रिशूल   ज्ञ माने ज्ञानी

1.1.1 Hindi Matra

	चिन्ह	संधि	उदाहरण
अ			कलम
आ कि मात्रा	ा	क + ा = का	कागज़
छोटी इ कि मात्रा	ि	क + ि = कि	किताब
बड़ी ई कि मात्रा	ी	क + ी = की	कीमत
छोटे उ कि मात्रा	ु	क + ु = कु	कुसुम
बड़े ऊ कि मात्रा	ू	क + ू = कू	कूदना
ऋ कि मात्रा	ृ	क + ृ = कृ	कृष्ण
ए कि मात्रा	े	क + े = के	केला
ऐ कि मात्रा	ै	क + ै = कै	कैरी
ओ कि मात्रा	ो	क + ो = को	कोयल
औ कि मात्रा	ौ	क + ौ = कौ	कौआ
अं कि मात्रा	ं	क + ं = कं	कंगन
अः कि मात्रा	ः	क + ः = कः न + ः = नः	पुनः

1.2 English Alphabets

A for Apple

B for Bus

C for Car

D for Dog

E for Elephant

F for Fish

G for Goat

H for Horse

I for Inkpot

J for Jug

K for Kite

L for Lotus

M for Moon

N for Nest

O for Orange

P for Peacock

Q for Queen

R for Rose

S for Ship

T for Tiger

U for Umbrella

V for Violin

W for Window

X for X-mas Tree

Y for Yak

Z for Zebra

Counting

1	One	★
2	Two	★ ★
3	Three	★ ★ ★
4	Four	★ ★ ★ ★
5	Five	★ ★ ★ ★ ★
6	Six	★ ★ ★ ★ ★ ★
7	Seven	★ ★ ★ ★ ★ ★ ★
8	Eight	★ ★ ★ ★ ★ ★ ★ ★
9	Nine	★ ★ ★ ★ ★ ★ ★ ★ ★
10	Ten	★ ★ ★ ★ ★ ★ ★ ★ ★ ★

Multiplication Tables

3.1 Hindi Tables

०	X	१	=	०
०	X	२	=	०
०	X	३	=	०
०	X	४	=	०
०	X	५	=	०
०	X	६	=	०
०	X	७	=	०
०	X	८	=	०
०	X	९	=	०
०	X	१0	=	०

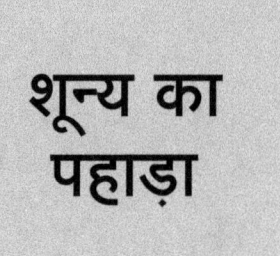

शून्य	एकम	शून्य
शून्य	दूनी	शून्य
शून्य	तीया	शून्य
शून्य	चोके	शून्य
शून्य	पंजे	शून्य
शून्य	छके	शून्य
शून्य	सत्ते	शून्य
शून्य	अठ्ठे	शून्य
शून्य	नौवे	शून्य
शून्य	दहाई	शून्य

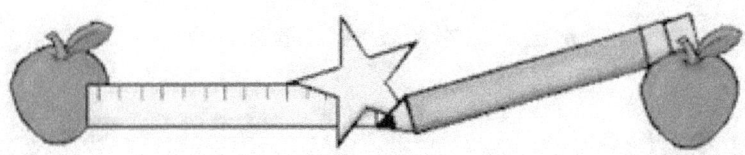

१	X	२	=	२
१	X	३	=	३
१	X	४	=	४
१	X	५	=	५
१	X	६	=	६
१	X	७	=	७
१	X	८	=	८
१	X	९	=	९
१	X	१0	=	१0

एक का पहाड़ा

एक	एकम	एक
एक	दूनी	दो
एक	तीया	तीन
एक	चोके	चार
एक	पंजे	पांच
एक	छके	छै
एक	सत्ते	सात
एक	अठ्ठे	आठ
एक	नौवे	नौ
एक	दहाई	दस

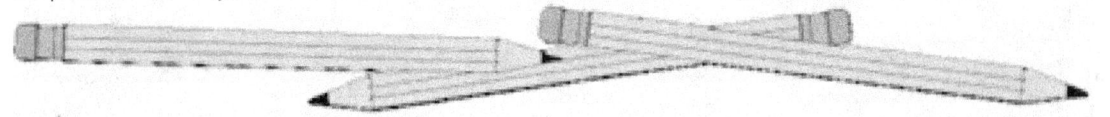

दो का पहाड़ा

२	×	१	=	२
२	×	२	=	४
२	×	३	=	६
२	×	४	=	८
२	×	५	=	१०
२	×	६	=	१२
२	×	७	=	१४
२	×	८	=	१६
२	×	९	=	१८
२	×	१०	=	२०

दो	एकम	दो
दो	एकम	दो
दो	दूनी	चार
दो	तीया	छै
दो	चोके	आठ
दो	पंजे	दस
दो	छक्के	बारह
दो	सत्ते	चौदह
दो	अठ्ठे	सोलह
दो	नौवे	अट्ठारह
दो	दहाई	बीस

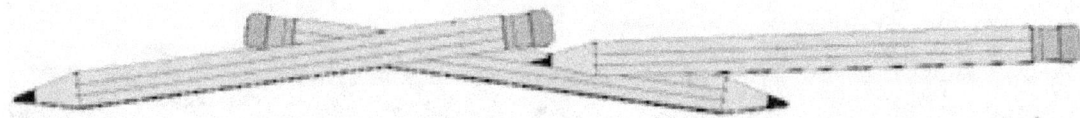

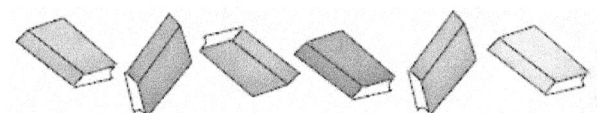

३	X	१	=	३
३	X	२	=	६
३	X	३	=	९
३	X	४	=	१२
३	X	५	=	१५
३	X	६	=	१८
३	X	७	=	२१
३	X	८	=	२४
३	X	९	=	२७
३	X	१०	=	३०

तीन का पहाड़ा

तीन	एकम	तीन
तीन	दूनी	छै
तीन	तीया	नौ
तीन	चोके	बारह
तीन	पंजे	पंद्रह
तीन	छके	अट्ठारह
तीन	सत्ते	इक्कीस
तीन	अठ्ठे	चौबिस
तीन	नौवे	सत्ताईस
तीन	दहाई	तीस

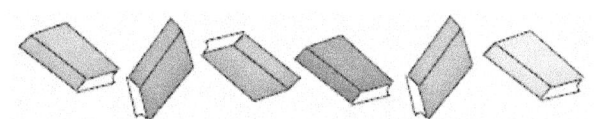

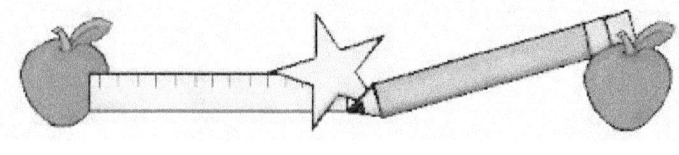

४	X	१	=	४
४	X	२	=	८
४	X	३	=	१२
४	X	४	=	१६
४	X	५	=	20
४	X	६	=	२४
४	X	७	=	२८
४	X	८	=	३२
४	X	९	=	३६
४	X	90	=	४0

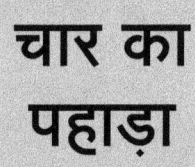

चार का पहाड़ा

चार	एकम	चार
चार	दूनी	आठ
चार	तीया	बारह
चार	चोके	सोलह
चार	पंजे	बीस
चार	छके	चौबिस
चार	सत्ते	अड्डाईस
चार	अठठे	बत्तीस
चार	नौवे	छत्तीस
चार	दहाई	चालीस

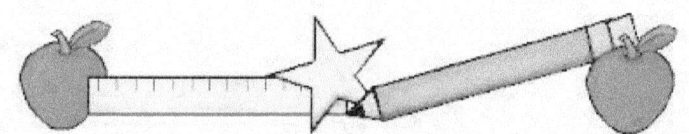

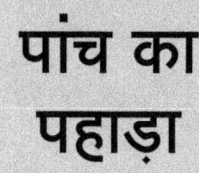

पांच का पहाड़ा

५	X	१	=	५
५	X	२	=	१०
५	X	३	=	१५
५	X	४	=	२०
५	X	५	=	२५
५	X	६	=	३०
५	X	७	=	३५
५	X	८	=	४०
५	X	९	=	४५
५	X	१०	=	५०

पांच	एकम	पांच
पांच	दूनी	दस
पांच	तीया	पंद्रह
पांच	चोके	बीस
पांच	पंजे	पच्चीस
पांच	छके	तीस
पांच	सत्ते	पैंतीस
पांच	अठ्ठे	चालीस
पांच	नौवे	पैंतालीस
पांच	दहाई	पचास

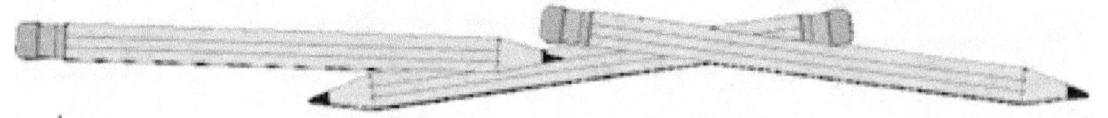

६	×	१	=	६
६	×	२	=	१२
६	×	३	=	१८
६	×	४	=	२४
६	×	५	=	३०
६	×	६	=	३६
६	×	७	=	४२
६	×	८	=	४८
६	×	९	=	५४
६	×	१०	=	६०

छै का पहाड़ा

छै	एकम	छै
छै	दूनी	बारह
छै	तीया	अट्ठारह
छै	चोके	चौबिस
छै	पंजे	तीस
छै	छके	छत्तीस
छै	सत्ते	बयालीस
छै	अठ्ठे	अइतालीस
छै	नौवे	चौवन
छै	दहाई	साठ

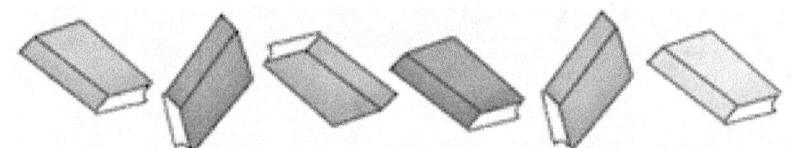

७	X	१	=	७
७	X	२	=	१४
७	X	३	=	२१
७	X	४	=	२८
७	X	५	=	३५
७	X	६	=	४२
७	X	७	=	४९
७	X	८	=	५६
७	X	९	=	६३
७	X	१०	=	७०

सात का पहाड़ा

सात	एकम	सात
सात	दूनी	चौदह
सात	तीया	इक्कीस
सात	चोके	अट्ठाईस
सात	पंजे	पैंतीस
सात	छके	बयालीस
सात	सत्ते	उनचास
सात	अठठे	छप्पन
सात	नौवे	तिरेसठ
सात	दहाई	सत्तर

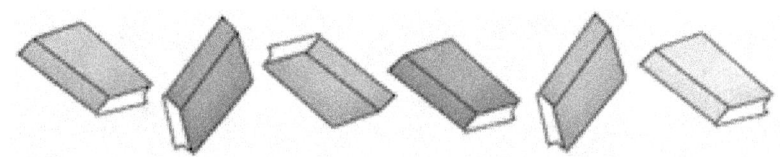

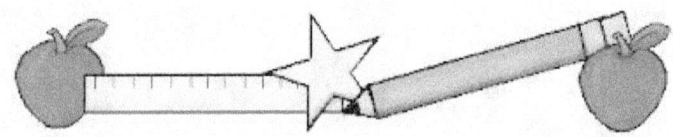

८	X	१	=	८
८	X	२	=	१६
८	X	३	=	२४
८	X	४	=	३२
८	X	५	=	४०
८	X	६	=	४८
८	X	७	=	५६
८	X	८	=	६४
८	X	९	=	७२
८	X	१०	=	८०

आठ का पहाड़ा

आठ	एकम	आठ
आठ	दूनी	सोलह
आठ	तीया	चौबिस
आठ	चोके	बत्तीस
आठ	पंजे	चालीस
आठ	छके	अड़तालीस
आठ	सत्ते	छप्पन
आठ	अठ्ठे	चौंसठ
आठ	नौवे	बहत्तर
आठ	दहाई	अस्सी

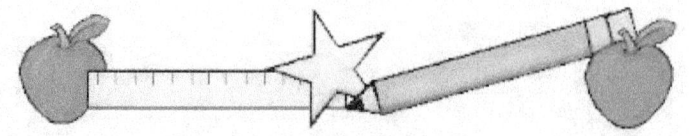

नौ का पहाड़ा

६	×	१	=	६
६	×	२	=	१२
६	×	३	=	१८
६	×	४	=	२४
६	×	५	=	३०
६	×	६	=	३६
६	×	७	=	४२
६	×	८	=	४८
६	×	९	=	५४
६	×	१0	=	६0

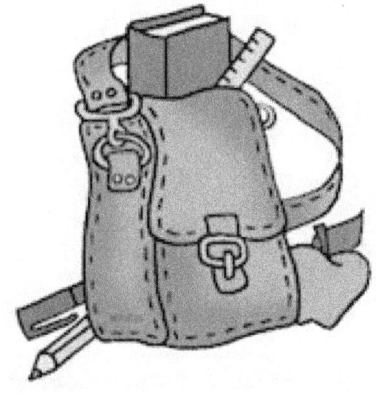

नौ	एकम	नौ
नौ	दूनी	अट्ठारह
नौ	तीया	सत्ताईस
नौ	चौके	छत्तीस
नौ	पंजे	पैंतालीस
नौ	छके	चौवन
नौ	सत्ते	तिरेसठ
नौ	अठ्ठे	बहत्तर
नौ	नौवे	इक्यासी
नौ	दहाई	नब्बे

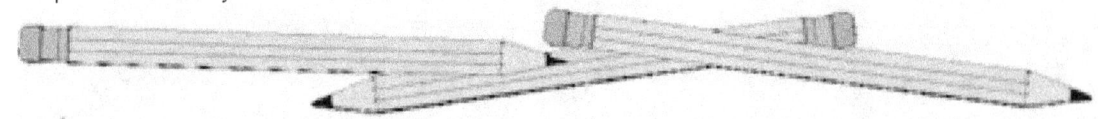

१०	X	१	=	१०
१०	X	२	=	२०
१०	X	३	=	३०
१०	X	४	=	४०
१०	X	५	=	५०
१०	X	६	=	६०
१०	X	७	=	७०
१०	X	८	=	८०
१०	X	९	=	९०
१०	X	१०	=	१००

दस का पहाड़ा

दस	एकम	दस
दस	दूनी	बीस
दस	तीया	तीस
दस	चोके	चालीस
दस	पंजे	पचास
दस	छके	साठ
दस	सत्ते	सत्तर
दस	अठ्ठे	अस्सी
दस	नौवे	नब्बे
दस	दहाई	सौ

११	X	१	=	११
११	X	२	=	२२
११	X	३	=	३३
११	X	४	=	४४
११	X	५	=	५५
११	X	६	=	६६
११	X	७	=	७७
११	X	८	=	८८
११	X	९	=	९९
११	X	१०	=	११०

ग्यारह का पहाड़ा

ग्यारह	एकम	ग्यारह
ग्यारह	दूनी	बाईस
ग्यारह	तीया	तैंतीस
ग्यारह	चोके	चौंतालीस
ग्यारह	पंजे	पचपन
ग्यारह	छके	छयासठ
ग्यारह	सत्ते	सतहत्तर
ग्यारह	अठ्ठे	अठासी
ग्यारह	नौवे	निन्यानवे
ग्यारह	दहाई	एक सौ दस

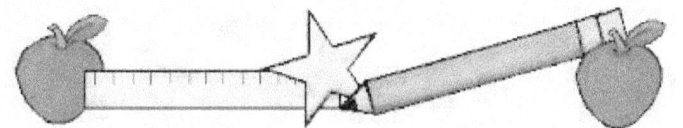

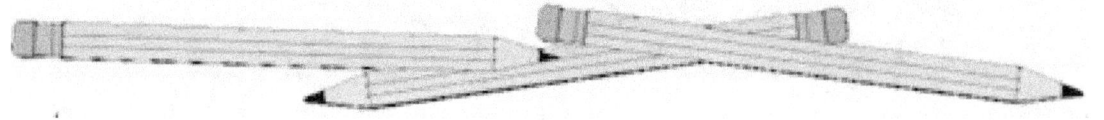

१२	X	१	=	१२
१२	X	२	=	२४
१२	X	३	=	३६
१२	X	४	=	४८
१२	X	५	=	६०
१२	X	६	=	७२
१२	X	७	=	८४
१२	X	८	=	९६
१२	X	९	=	१०८
१२	X	१०	=	१२०

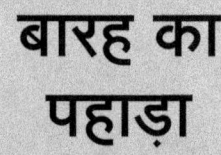

बारह का पहाड़ा

बारह	एकम	बारह
बारह	दूनी	चौबीस
बारह	तीया	छत्तीस
बारह	चोके	अड़तालीस
बारह	पंजे	साठ
बारह	छके	बहत्तर
बारह	सत्ते	चौरासी
बारह	अठ्ठे	छियान्वे
बारह	नौवे	एक सौ आठ
बारह	दहाई	एक सौ बीस

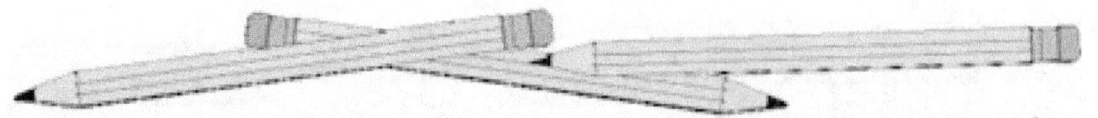

१३	X	१	=	१३
१३	X	२	=	२६
१३	X	३	=	३९
१३	X	४	=	५२
१३	X	५	=	६५
१३	X	६	=	७८
१३	X	७	=	९१
१३	X	८	=	१०४
१३	X	९	=	११७
१३	X	१०	=	१३०

तेरह का पहाड़ा

तेरह	एकम	तेरह
तेरह	दूनी	छब्बीस
तेरह	तीया	उनतालीस
तेरह	चोके	बावन
तेरह	पंजे	पैंसठ
तेरह	छके	अठहत्तर
तेरह	सत्ते	इक्यानवे
तेरह	अठ्ठे	एक सौ चार
तेरह	नौवे	एक सौ सत्रह
तेरह	दहाई	एक सौ तीस

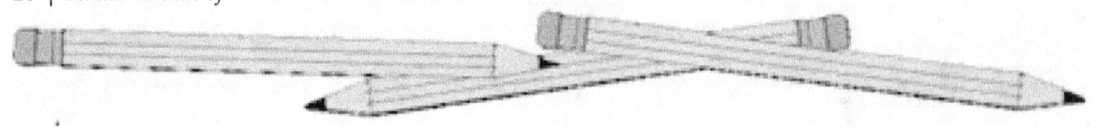

१४	x	१	=	१४
१४	x	२	=	२८
१४	x	३	=	४२
१४	x	४	=	५६
१४	x	५	=	७०
१४	x	६	=	८४
१४	x	७	=	९८
१४	x	८	=	११२
१४	x	९	=	१२६
१४	x	१०	=	१४०

चौदह का पहाड़ा

चौदह	एकम	चौदह
चौदह	दूनी	अट्ठाईस
चौदह	तीया	बयालीस
चौदह	चोके	छप्पन
चौदह	पंजे	सत्तर
चौदह	छके	चौरासी
चौदह	सत्ते	अट्ठानवे
चौदह	अठ्ठे	एक सौ बारह
चौदह	नौवे	एक सौ छब्बीस
चौदह	दहाई	एक सौ चालीस

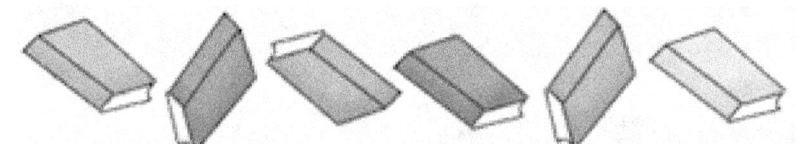

१५	×	१	=	१५
१५	×	२	=	३०
१५	×	३	=	४५
१५	×	४	=	६०
१५	×	५	=	७५
१५	×	६	=	९०
१५	×	७	=	१०५
१५	×	८	=	१२०
१५	×	९	=	१३५
१५	×	१०	=	१५०

पंद्रह का पहाड़ा

पंद्रह	एकम	पंद्रह
पंद्रह	दूनी	तीस
पंद्रह	तीया	पैंतालीस
पंद्रह	चोके	साठ
पंद्रह	पंजे	पचहत्तर
पंद्रह	छके	नब्बे
पंद्रह	सत्ते	एक सौ पांच
पंद्रह	अठ्ठे	एक सौ बीस
पंद्रह	नौवे	एक सौ पैंतीस
पंद्रह	दहाई	एक सौ पचास

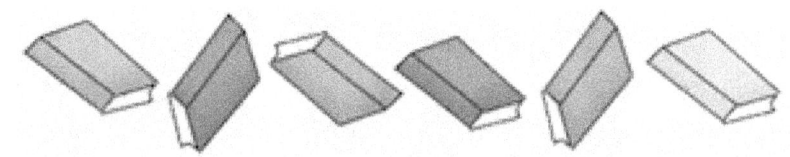

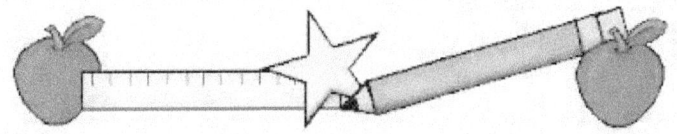

१६	X	१	=	१६
१६	X	२	=	३२
१६	X	३	=	४८
१६	X	४	=	६४
१६	X	५	=	८०
१६	X	६	=	९६
१६	X	७	=	११२
१६	X	८	=	१२८
१६	X	९	=	१४४
१६	X	१०	=	१६०

सोलह का पहाड़ा

सोलह	एकम	सोलह
सोलह	दूनी	बत्तीस
सोलह	तीया	अड़तालीस
सोलह	चोके	चौंसठ
सोलह	पंजे	अस्सी
सोलह	छके	छियानवे
सोलह	सत्ते	एक सौ बारह
सोलह	अठ्ठे	एक सौ अड्डाईस
सोलह	नौवे	एक सौ चौंतालीस
सोलह	दहाई	एक सौ साठ

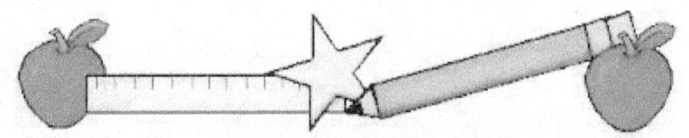

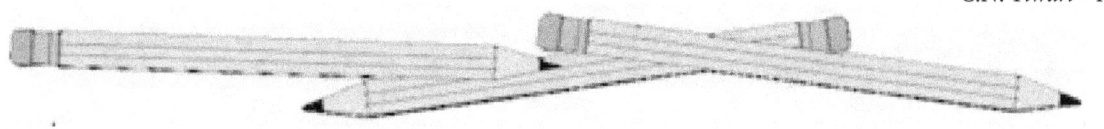

१७	X	१	=	१७
१७	X	२	=	३४
१७	X	३	=	५१
१७	X	४	=	६८
१७	X	५	=	८५
१७	X	६	=	१०२
१७	X	७	=	११९
१७	X	८	=	१३६
१७	X	९	=	१५३
१७	X	१०	=	१७०

सत्रह का पहाड़ा

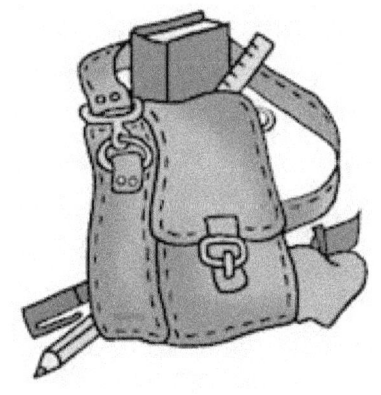

सत्रह	एकम	सत्रह
सत्रह	दूनी	चौंतिस
सत्रह	तीया	इक्यावन
सत्रह	चोके	अडसठ
सत्रह	पंजे	पचासी
सत्रह	छके	एक सौ दस
सत्रह	सत्ते	एक सौ उन्नीस
सत्रह	अठ्ठे	एक सौ छत्तीस
सत्रह	नौवे	एक सौ तिरेपन
सत्रह	दहाई	एक सौ सत्तर

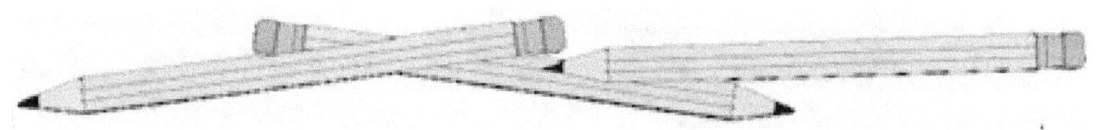

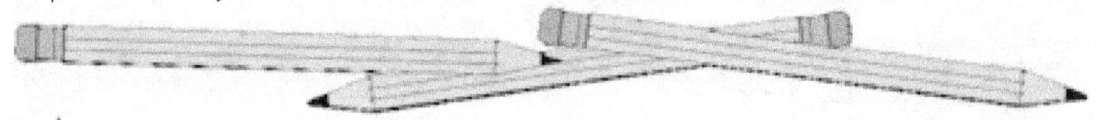

१८	x	१	=	१८
१८	x	२	=	३६
१८	x	३	=	५४
१८	x	४	=	७२
१८	x	५	=	९०
१८	x	६	=	१०८
१८	x	७	=	१२६
१८	x	८	=	१४४
१८	x	९	=	१६२
१८	x	१०	=	१८०

अट्ठारह का पहाड़ा

अट्ठारह	एकम	अट्ठारह
अट्ठारह	दूनी	छत्तीस
अट्ठारह	तीया	चौवन
अट्ठारह	चोके	बहत्तर
अट्ठारह	पंजे	नब्बे
अट्ठारह	छके	एक सौ आठ
अट्ठारह	सत्ते	एक सौ छब्बीस
अट्ठारह	अठ्ठे	एक सौ चौंतालीस
अट्ठारह	नौवे	एक सो बासठ
अट्ठारह	दहाई	एक सौ अस्सी

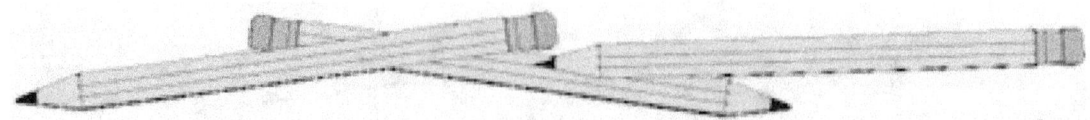

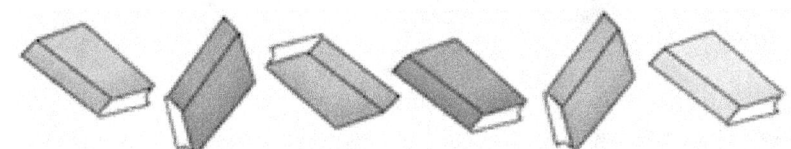

१९	×	१	=	१९
१९	×	२	=	३८
१९	×	३	=	५७
१९	×	४	=	७६
१९	×	५	=	९५
१९	×	६	=	११४
१९	×	७	=	१३३
१९	×	८	=	१५२
१९	×	९	=	१७१
१९	×	१०	=	१९०

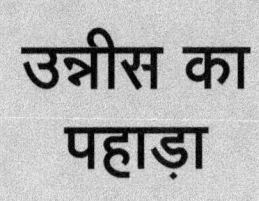

उन्नीस का पहाड़ा

उन्नीस	एकम	उन्नीस
उन्नीस	दूनी	अड़तीस
उन्नीस	तीया	सत्तावन
उन्नीस	चोके	छिहत्तर
उन्नीस	पंजे	पचानवे
उन्नीस	छके	एक सौ चौदह
उन्नीस	सत्ते	एक सौ तैंतीस
उन्नीस	अठ्ठे	एक सौ बावन
उन्नीस	नौवे	एक सौ इकहत्तर
उन्नीस	दहाई	एक सौ नब्बे

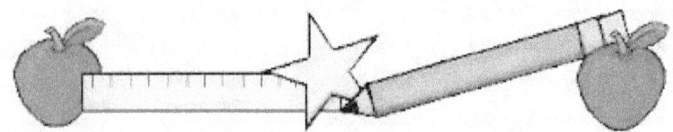

२०	X	१	=	२०
२०	X	२	=	४०
२०	X	३	=	६०
२०	X	४	=	८०
२०	X	५	=	१००
२०	X	६	=	१२०
२०	X	७	=	१४०
२०	X	८	=	१६०
२०	X	९	=	१८०
२०	X	१०	=	२००

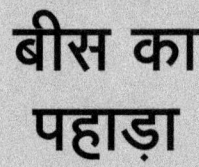

बीस का
पहाड़ा

बीस	एकम	बीस
बीस	दूनी	चालीस
बीस	तीया	साठ
बीस	चोके	अस्सी
बीस	पंजे	सौ
बीस	छके	एक सौ बीस
बीस	सत्ते	एक सौ चालीस
बीस	अठ्ठे	एक सौ साठ
बीस	नौवे	एक सौ अस्सी
बीस	दहाई	दो सौ

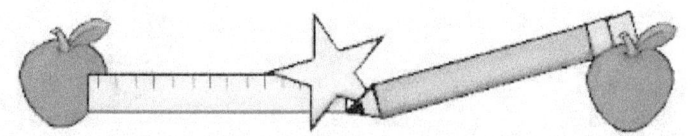

3.2 English Tables

I CAN READ LIKE THIS				I CAN ALSO READ LIKE THIS[1*]			
0	one	is	0	0	one	is	0
0	twos	are	0	0	two	sa	0
0	threes	are	0	0	three	sa	0
0	fours	are	0	0	four	sa	0
0	fives	are	0	0	five	sa	0
0	sixes	are	0	0	six	sa	0
0	sevens	are	0	0	seven	sa	0
0	eights	are	0	0	eight	sa	0
0	nines	are	0	0	nine	sa	0
0	tens	are	0	0	ten	sa	0

I CAN WRITE LIKE THIS

0	x	1	=	0
0	x	2	=	0
0	x	3	=	0
0	x	4	=	0
0	x	5	=	0
0	x	6	=	0
0	x	7	=	0
0	x	8	=	0
0	x	9	=	0
0	x	10	=	0

---

1 * We can also read it as '0 two za 0' and so on. According to British English, the sound of 's' in twos is z and 'are' is pronounced as aa. Therefore, upon merging twos and are, the resultant becomes 'two za'.

I CAN READ LIKE THIS				I CAN ALSO READ LIKE THIS			
1	one	is	1	1	one	is	1
1	twos	are	2	1	two	sa	2
1	threes	are	3	1	three	sa	3
1	fours	are	4	1	four	sa	4
1	fives	are	5	1	five	sa	5
1	sixes	are	6	1	six	sa	6
1	sevens	are	7	1	seven	sa	7
1	eights	are	8	1	eight	sa	8
1	nines	are	9	1	nine	sa	9
1	tens	are	10	1	ten	sa	10

I CAN WRITE LIKE THIS

1	x	1	=	1
1	x	2	=	2
1	x	3	=	3
1	x	4	=	4
1	x	5	=	5
1	x	6	=	6
1	x	7	=	7
1	x	8	=	8
1	x	9	=	9
1	x	10	=	10

I CAN READ LIKE THIS				I CAN ALSO READ LIKE THIS			
2	one	is	2	2	one	is	2
2	twos	are	4	2	two	sa	4
2	threes	are	6	2	three	sa	6
2	fours	are	8	2	four	sa	8
2	fives	are	10	2	five	sa	10
2	sixes	are	12	2	six	sa	12
2	sevens	are	14	2	seven	sa	14
2	eights	are	16	2	eight	sa	16
2	nines	are	18	2	nine	sa	18
2	tens	are	20	2	ten	sa	20

I CAN WRITE LIKE THIS

2	x	1	=	2
2	x	2	=	4
2	x	3	=	6
2	x	4	=	8
2	x	5	=	10
2	x	6	=	12
2	x	7	=	14
2	x	8	=	16
2	x	9	=	18
2	x	10	=	20

I CAN READ LIKE THIS				I CAN ALSO READ LIKE THIS			
3	one	is	3	3	one	is	3
3	twos	are	6	3	two	sa	6
3	threes	are	9	3	three	sa	9
3	fours	are	12	3	four	sa	12
3	fives	are	15	3	five	sa	15
3	sixes	are	18	3	six	sa	18
3	sevens	are	21	3	seven	sa	21
3	eights	are	24	3	eight	sa	24
3	nines	are	27	3	nine	sa	27
3	tens	are	30	3	ten	sa	30

I CAN WRITE LIKE THIS

3	x	1	=	3
3	x	2	=	6
3	x	3	=	9
3	x	4	=	12
3	x	5	=	15
3	x	6	=	18
3	x	7	=	21
3	x	8	=	24
3	x	9	=	27
3	x	10	=	30

I CAN READ LIKE THIS				I CAN ALSO READ LIKE THIS			
4	one	is	4	4	one	is	4
4	twos	are	8	4	two	sa	8
4	threes	are	12	4	three	sa	12
4	fours	are	16	4	four	sa	16
4	fives	are	20	4	five	sa	20
4	sixes	are	24	4	six	sa	24
4	sevens	are	28	4	seven	sa	28
4	eights	are	32	4	eight	sa	32
4	nines	are	36	4	nine	sa	36
4	tens	are	40	4	ten	sa	40

I CAN WRITE LIKE THIS

4	x	1	=	4
4	x	2	=	8
4	x	3	=	12
4	x	4	=	16
4	x	5	=	20
4	x	6	=	24
4	x	7	=	28
4	x	8	=	32
4	x	9	=	36
4	x	10	=	40

I CAN READ LIKE THIS				I CAN ALSO READ LIKE THIS			
5	one	is	5	5	one	is	5
5	twos	are	10	5	two	sa	10
5	threes	are	15	5	three	sa	15
5	fours	are	20	5	four	sa	20
5	fives	are	25	5	five	sa	25
5	sixes	are	30	5	six	sa	30
5	sevens	are	35	5	seven	sa	35
5	eights	are	40	5	eight	sa	40
5	nines	are	45	5	nine	sa	45
5	tens	are	50	5	ten	sa	50

I CAN WRITE LIKE THIS

5	x	1	=	5
5	x	2	=	10
5	x	3	=	15
5	x	4	=	20
5	x	5	=	25
5	x	6	=	30
5	x	7	=	35
5	x	8	=	40
5	x	9	=	45
5	x	10	=	50

I CAN READ LIKE THIS				I CAN ALSO READ LIKE THIS			
6	one	is	6	6	one	is	6
6	twos	are	12	6	two	sa	12
6	threes	are	18	6	three	sa	18
6	fours	are	24	6	four	sa	24
6	fives	are	30	6	five	sa	30
6	sixes	are	36	6	six	sa	36
6	sevens	are	42	6	seven	sa	42
6	eights	are	48	6	eight	sa	48
6	nines	are	54	6	nine	sa	54
6	tens	are	60	6	ten	sa	60

I CAN WRITE LIKE THIS

6	x	1	=	6
6	x	2	=	12
6	x	3	=	18
6	x	4	=	24
6	x	5	=	30
6	x	6	=	36
6	x	7	=	42
6	x	8	=	48
6	x	9	=	54
6	x	10	=	60

I CAN READ LIKE THIS				I CAN ALSO READ LIKE THIS			
7	one	is	7	7	one	is	7
7	twos	are	14	7	two	sa	14
7	threes	are	21	7	three	sa	21
7	fours	are	28	7	four	sa	28
7	fives	are	35	7	five	sa	35
7	sixes	are	42	7	six	sa	42
7	sevens	are	49	7	seven	sa	49
7	eights	are	56	7	eight	sa	56
7	nines	are	63	7	nine	sa	63
7	tens	are	70	7	ten	sa	70

I CAN WRITE LIKE THIS

7	x	1	=	7
7	x	2	=	14
7	x	3	=	21
7	x	4	=	28
7	x	5	=	35
7	x	6	=	42
7	x	7	=	49
7	x	8	=	56
7	x	9	=	63
7	x	10	=	70

I CAN READ LIKE THIS				I CAN ALSO READ LIKE THIS			
8	one	is	8	8	one	is	8
8	twos	are	16	8	two	sa	16
8	threes	are	24	8	three	sa	24
8	fours	are	32	8	four	sa	32
8	fives	are	40	8	five	sa	40
8	sixes	are	48	8	six	sa	48
8	sevens	are	56	8	seven	sa	56
8	eights	are	64	8	eight	sa	64
8	nines	are	72	8	nine	sa	72
8	tens	are	80	8	ten	sa	80

I CAN WRITE LIKE THIS

8	x	1	=	8
8	x	2	=	16
8	x	3	=	24
8	x	4	=	32
8	x	5	=	40
8	x	6	=	48
8	x	7	=	56
8	x	8	=	64
8	x	9	=	72
8	x	10	=	80

I CAN READ LIKE THIS				I CAN ALSO READ LIKE THIS			
9	one	is	9	9	one	is	9
9	twos	are	18	9	two	sa	18
9	threes	are	27	9	three	sa	27
9	fours	are	36	9	four	sa	36
9	fives	are	45	9	five	sa	45
9	sixes	are	54	9	six	sa	54
9	sevens	are	63	9	seven	sa	63
9	eights	are	73	9	eight	sa	72
9	nines	are	81	9	nine	sa	81
9	tens	are	90	9	ten	sa	90

I CAN WRITE LIKE THIS

9	x	1	=	9
9	x	2	=	18
9	x	3	=	27
9	x	4	=	36
9	x	5	=	45
9	x	6	=	54
9	x	7	=	63
9	x	8	=	72
9	x	9	=	81
9	x	10	=	90

I CAN READ LIKE THIS				I CAN ALSO READ LIKE THIS			
10	one	is	10	10	one	is	10
10	twos	are	20	10	two	sa	20
10	threes	are	30	10	three	sa	30
10	fours	are	40	10	four	sa	40
10	fives	are	50	10	five	sa	50
10	sixes	are	60	10	six	sa	60
10	sevens	are	70	10	seven	sa	70
10	eights	are	80	10	eight	sa	80
10	nines	are	90	10	nine	sa	90
10	tens	are	100	10	ten	sa	100

I CAN WRITE LIKE THIS

10	x	1	=	10
10	x	2	=	20
10	x	3	=	30
10	x	4	=	40
10	x	5	=	50
10	x	6	=	60
10	x	7	=	70
10	x	8	=	80
10	x	9	=	90
10	x	10	=	100

I CAN READ LIKE THIS				I CAN ALSO READ LIKE THIS			
11	one	is	11	11	one	is	11
11	twos	are	22	11	two	sa	22
11	threes	are	33	11	three	sa	33
11	fours	are	44	11	four	sa	44
11	fives	are	55	11	five	sa	55
11	sixes	are	66	11	six	sa	66
11	sevens	are	77	11	seven	sa	77
11	eights	are	88	11	eight	sa	88
11	nines	are	99	11	nine	sa	99
11	tens	are	110	11	ten	sa	110

I CAN WRITE LIKE THIS

11	x	1	=	11
11	x	2	=	22
11	x	3	=	33
11	x	4	=	44
11	x	5	=	55
11	x	6	=	66
11	x	7	=	77
11	x	8	=	88
11	x	9	=	99
11	x	10	=	110

I CAN READ LIKE THIS				I CAN ALSO READ LIKE THIS			
12	one	is	12	12	one	is	12
12	twos	are	24	12	two	sa	24
12	threes	are	36	12	three	sa	36
12	fours	are	48	12	four	sa	48
12	fives	are	60	12	five	sa	60
12	sixes	are	72	12	six	sa	72
12	sevens	are	84	12	seven	sa	84
12	eights	are	96	12	eight	sa	96
12	nines	are	108	12	nine	sa	108
12	tens	are	120	12	ten	sa	120

I CAN WRITE LIKE THIS

12	x	1	=	12
12	x	2	=	24
12	x	3	=	36
12	x	4	=	48
12	x	5	=	60
12	x	6	=	72
12	x	7	=	84
12	x	8	=	96
12	x	9	=	108
12	x	10	=	120

I CAN READ LIKE THIS				I CAN ALSO READ LIKE THIS			
13	one	is	13	13	one	is	13
13	twos	are	26	13	two	sa	26
13	threes	are	39	13	three	sa	39
13	fours	are	52	13	four	sa	52
13	fives	are	65	13	five	sa	65
13	sixes	are	78	13	six	sa	78
13	sevens	are	91	13	seven	sa	91
13	eights	are	104	13	eight	sa	104
13	nines	are	117	13	nine	sa	117
13	tens	are	130	13	ten	sa	130

I CAN WRITE LIKE THIS

13	x	1	=	13
13	x	2	=	26
13	x	3	=	39
13	x	4	=	52
13	x	5	=	65
13	x	6	=	78
13	x	7	=	91
13	x	8	=	104
13	x	9	=	117
13	x	10	=	130

I CAN READ LIKE THIS			
14	one	is	14
14	twos	are	28
14	threes	are	42
14	fours	are	56
14	fives	are	70
14	sixes	are	84
14	sevens	are	98
14	eights	are	112
14	nines	are	126
14	tens	are	140

I CAN ALSO READ LIKE THIS			
14	one	is	14
14	two	sa	28
14	three	sa	42
14	four	sa	56
14	five	sa	70
14	six	sa	84
14	seven	sa	98
14	eight	sa	112
14	nine	sa	126
14	ten	sa	140

I CAN WRITE LIKE THIS

14	x	1	=	14
14	x	2	=	28
14	x	3	=	42
14	x	4	=	56
14	x	5	=	70
14	x	6	=	84
14	x	7	=	98
14	x	8	=	112
14	x	9	=	126
14	x	10	=	140

I CAN READ LIKE THIS				I CAN ALSO READ LIKE THIS			
15	one	is	15	15	one	is	15
15	twos	are	30	15	two	sa	30
15	threes	are	45	15	three	sa	45
15	fours	are	60	15	four	sa	60
15	fives	are	75	15	five	sa	75
15	sixes	are	90	15	six	sa	90
15	sevens	are	105	15	seven	sa	105
15	eights	are	120	15	eight	sa	120
15	nines	are	135	15	nine	sa	135
15	tens	are	150	15	ten	sa	150

I CAN WRITE LIKE THIS

15	x	1	=	15
15	x	2	=	30
15	x	3	=	45
15	x	4	=	60
15	x	5	=	75
15	x	6	=	90
15	x	7	=	105
15	x	8	=	120
15	x	9	=	135
15	x	10	=	150

I CAN READ LIKE THIS				I CAN ALSO READ LIKE THIS			
16	one	is	16	16	one	is	16
16	twos	are	32	16	two	sa	32
16	threes	are	48	16	three	sa	48
16	fours	are	64	16	four	sa	64
16	fives	are	80	16	five	sa	80
16	sixes	are	96	16	six	sa	96
16	sevens	are	112	16	seven	sa	112
16	eights	are	128	16	eight	sa	128
16	nines	are	144	16	nine	sa	144
16	tens	are	160	16	ten	sa	160

I CAN WRITE LIKE THIS

16	x	1	=	16
16	x	2	=	32
16	x	3	=	48
16	x	4	=	64
16	x	5	=	80
16	x	6	=	96
16	x	7	=	112
16	x	8	=	128
16	x	9	=	144
16	x	10	=	160

I CAN READ LIKE THIS				I CAN ALSO READ LIKE THIS			
17	one	is	17	17	one	is	17
17	twos	are	34	17	two	sa	34
17	threes	are	51	17	three	sa	51
17	fours	are	68	17	four	sa	68
17	fives	are	85	17	five	sa	85
17	sixes	are	102	17	six	sa	102
17	sevens	are	119	17	seven	sa	119
17	eights	are	136	17	eight	sa	136
17	nines	are	153	17	nine	sa	153
17	tens	are	170	17	ten	sa	170

I CAN WRITE LIKE THIS

17	x	1	=	17
17	x	2	=	34
17	x	3	=	51
17	x	4	=	68
17	x	5	=	85
17	x	6	=	102
17	x	7	=	119
17	x	8	=	136
17	x	9	=	153
17	x	10	=	170

I CAN READ LIKE THIS				I CAN ALSO READ LIKE THIS			
18	one	is	18	18	one	is	18
18	twos	are	36	18	two	sa	36
18	threes	are	54	18	three	sa	54
18	fours	are	72	18	four	sa	72
18	fives	are	90	18	five	sa	90
18	sixes	are	108	18	six	sa	108
18	sevens	are	126	18	seven	sa	126
18	eights	are	144	18	eight	sa	144
18	nines	are	162	18	nine	sa	162
18	tens	are	180	18	ten	sa	180

I CAN WRITE LIKE THIS

18	x	1	=	18
18	x	2	=	36
18	x	3	=	54
18	x	4	=	72
18	x	5	=	90
18	x	6	=	108
18	x	7	=	126
18	x	8	=	144
18	x	9	=	162
18	x	10	=	180

I CAN READ LIKE THIS				I CAN ALSO READ LIKE THIS			
19	one	is	19	19	one	is	19
19	twos	are	38	19	two	sa	38
19	threes	are	57	19	three	sa	57
19	fours	are	76	19	four	sa	76
19	fives	are	95	19	five	sa	95
19	sixes	are	114	19	six	sa	114
19	sevens	are	133	19	seven	sa	133
19	eights	are	152	19	eight	sa	152
19	nines	are	171	19	nine	sa	171
19	tens	are	190	19	ten	sa	190

I CAN WRITE LIKE THIS

19	x	1	=	19
19	x	2	=	38
19	x	3	=	57
19	x	4	=	76
19	x	5	=	95
19	x	6	=	114
19	x	7	=	133
19	x	8	=	152
19	x	9	=	171
19	x	10	=	190

I CAN READ LIKE THIS				I CAN ALSO READ LIKE THIS			
20	one	is	20	20	one	is	20
20	twos	are	40	20	two	sa	40
20	threes	are	60	20	three	sa	60
20	fours	are	80	20	four	sa	80
20	fives	are	100	20	five	sa	100
20	sixes	are	120	20	six	sa	120
20	sevens	are	140	20	seven	sa	140
20	eights	are	160	20	eight	sa	160
20	nines	are	180	20	nine	sa	180
20	tens	are	200	20	ten	sa	200

I CAN WRITE LIKE THIS

20	x	1	=	20
20	x	2	=	40
20	x	3	=	60
20	x	4	=	80
20	x	5	=	100
20	x	6	=	120
20	x	7	=	140
20	x	8	=	160
20	x	9	=	180
20	x	10	=	200

Number Theory

4.1 Natural Numbers

Natural numbers denote the common way of counting (hence the term 'natural'). The numbers **1, 2, 3, 4, 5, 6, 7, 8, 9 and so on** are called natural numbers. Natural numbers are positive numbers. **A number without any sign is considered positive**. For example: 5 = +5 , is a natural number. Similarly, 1678 = +1678, is also a natural number.

4.2 Whole Numbers

The idea of **zero**, though natural to us now, was not natural to early humans. Zero was later added to the set of counting numbers to make a new set called whole numbers. The numbers **0, 1, 2, 3, 4, 5, 6, 7, 8, 9 and so on** are known as whole numbers.

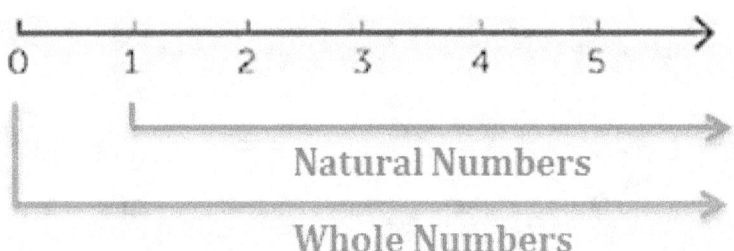

4.3 Negative Numbers

When we count forward we get natural numbers and whole numbers. Similarly, if we count backward we will get negative numbers such as -1, -2, -3, and so on. Note that 0 is neither positive nor negative.

A number less than 0 is called a negative number. Example: the temperature – 20° C is 20° below zero.

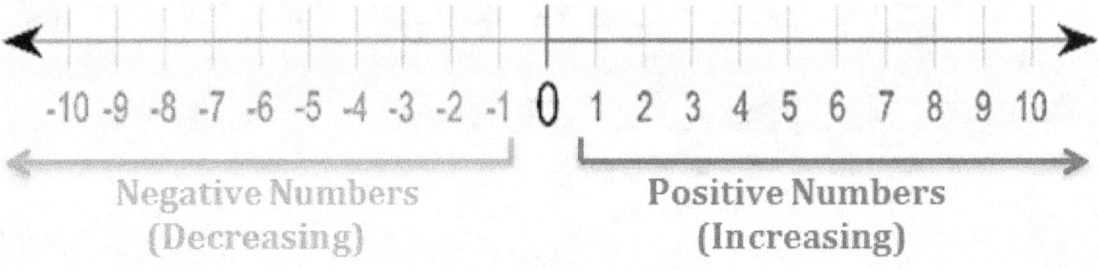

4.4 Integers

If we combine negative numbers with the whole numbers, we get a new set of numbers termed as Integers.

Integers: {. . ., –4, –3, –2, –1, 0, 1, 2, 3, 4,. . .}

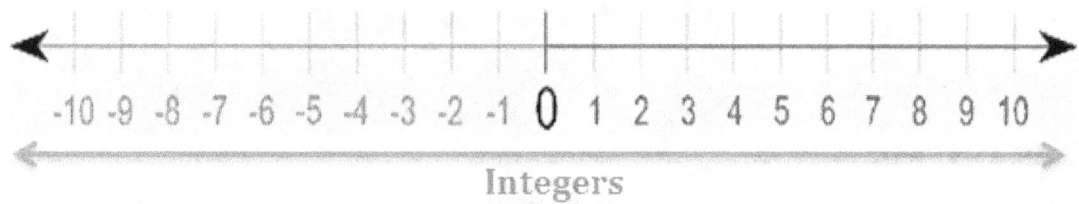

4.5 Operation

An operation is a mathematical process. The most common operations are addition, subtraction, multiplication and division. Each of these four basic operations is described next.

4.5.1 Addition

Addition is finding the total or sum by combining two or more numbers. The symbol for addition is + (read as plus). Addition of two numbers is represented as:

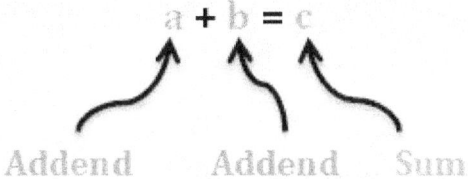

Example:

2 + 6 = 8

Here, 2 and 6 are **addend** and 8 is the **sum**.

Properties:

1. **Commutative Law: a + b = b + a**
   For example, 4 + 6 = 6 + 4 = 10
2. **Associative Law: (a + b) + c = a + (b + c)**
   For example, (2 + 3) + 6 = 2 + (3 + 6) = 11
3. Sum of two positive numbers is a positive number.
   For example, 2 + 5 = 7

$$
\begin{array}{r}
+2 \\
+\ +5 \\
\hline
+7
\end{array}
$$

4. Sum of two negative numbers is a negative number.

   For example, $(-2) + (-5) = -7$

$$\begin{array}{r} -2 \\ +\ -5 \\ \hline -7 \\ \hline \end{array}$$

5. Sum of a positive and a negative number is equivalent to subtraction.

4.5.2 Subtraction

The operation of taking one number away from another is termed as subtraction. For example, if you have 5 oranges and you subtract 2, you are left with 3 oranges.

The symbol for subtraction is – (read as minus). Subtraction is represented as:

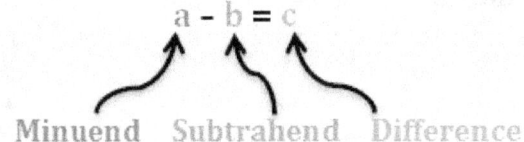

Minuend   Subtrahend   Difference

Minuend: The number that is to be subtracted from.

Subtrahend: The number that is to be subtracted.

Difference: The result of subtracting subtrahend from minuend.

Example:

$8 - 2 = 6$

Here, 8 is the **minuend**, 2 is the **subtrahend** and 6 is the **difference**.

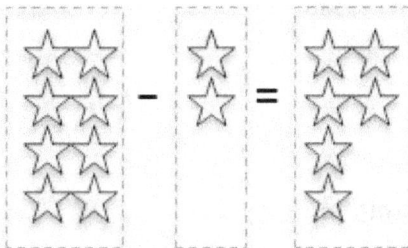

Properties:

1. Subtraction is not commutative. $a - b \neq b - a$
2. Subtraction is not associative. $(a - b) - c \neq a - (b - c)$
3. Difference between two numbers is **positive** if the **larger number is positive**. For example, $5 - 3 = +2$

$$\begin{array}{r} 5 \\ -\ \ 3 \\ \hline +2 \\ \hline \end{array}$$

Difference between two numbers is **negative** if the **larger number is negative**. For example, $3 - 5 = -2$

4.5.3 Multiplication

Multiplication is an operation of adding a number to itself, a specified number of times. The number being repeatedly added is known as **multiplicand** and the number of times it is added is known as **multiplier**.

The symbol for multiplication is × (read as times or multiplied by).

Multiplication is represented as:

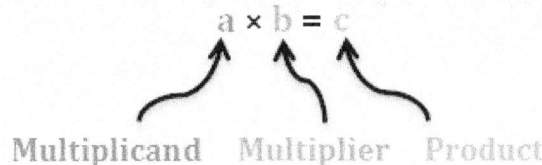

Example:

$$3 \times 4 = 3 + 3 + 3 + 3 = 12$$

4 times

Here, 3 is the **multiplicand**, 4 is the **multiplier** and 12 is the **product**.

Properties:

1. **Commutative Law: a × b = b × a**
   For example, 2 × 6 = 6 × 2 = 12
2. **Associative Law: (a × b) × c = a × (b × c)**
   For example, (2 × 3) × 6 = 2 × (3 × 6) = 36
3. The sign of the product of two numbers is obtained as:

$$+ \times + = +$$
$$+ \times - = -$$
$$- \times + = -$$
$$- \times - = +$$

   If the **signs** are **same** (both positive or both negative), then the product is **positive**.
   If the **signs** are **different**, then the product is **negative**.
4. a × b × c = b × c × a = c × b × a
   3 × 5 × 2 = 5 × 2 × 3 = 2 × 5 × 3 = 30
5. For any number a; a × 1 = 1 × a = a
   5 × 1 = 5; 10 × 1 = 10; 1000 × 1 = 1000
6. For any number a; a × 0 = 0 × a = 0
   5 × 0 = 0; 10 × 0 = 0; 1000 × 0 = 0
7. Multiplication by powers of 10:

   3 × 10 = 30   5 × 10 = 50
   3 × 100 = 300   5 × 100 = 500
   3 × 1000 = 3000   5 × 1000 = 5000
   3 × 10000 = 30000   5 × 10000 = 50000

4.5.4 Division

Division is a sharing operation where objects are shared (or divided) into a number of groups of equal number. The symbol for division is ÷ (read as by or divided by).

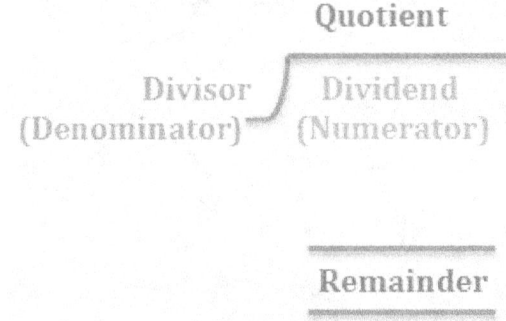

Example:

$$22 \div 2 = \frac{22}{2} = 11$$

Numerator ↑ (pointing to 22)

Denominator ↑ (pointing to 2)

| Dividend = Divisor × Quotient + Remainder |

$$22 = 2 \times 11 + 0$$

$$\begin{array}{r} 11 \\ 2\overline{\smash{)}22} \\ \underline{2{\downarrow}} \\ 02 \\ \underline{2} \\ 0 \end{array}$$

Properties:

1. Division is not commutative. $a \div b \neq b \div a$
2. Division is not associative. $(a \div b) \div c \neq a \div (b \div c)$
3. The sign of the quotient of two numbers is obtained as:

$$+ \div + = \frac{+}{+} = +$$

$$+ \div - = \frac{+}{-} = -$$

$$- \div + = \frac{-}{+} = -$$

$$- \div - = \frac{-}{-} = +$$

If the **signs** are **same** (both positive or both negative), then the quotient is **positive**.

If the **signs** are **different**, then the quotient is **negative**.

4. **Remainder is always positive.**

5. For any number a, $0 \div a = \dfrac{0}{a} = 0$

6. For any number a, $a \div 0 = \dfrac{a}{0} = \infty$

   (where, ∞ is infinity, a quantity which is bigger than any number. The value is of ∞ is undefined)

7. For any number a, $\infty \div a = \dfrac{a}{\infty} = \infty$

8. For any number a, $a \div \infty = \dfrac{\infty}{a} = 0$

Note: Irrespective of the operation (addition, subtraction, multiplication or division), first check the sign. Use the properties of each operation to deduce the sign of the result first. Only after the sign has been determined, solve for the numerical value of the result.

4.6 Exponent

The exponent or power of a number denotes how many times to use that number in a multiplication. For example: In 6^2, '2' is called the exponent or power of the number '6', and the power of '2' means we have to multiply '6' two times.

$$6^2 = 6 \times 6 = 36$$

6^2 can be read as "6 to the power of 2" or "6 to the second power" or simply "6 squared."

Note: The exponent or power of '2' of a number is also called **square** of that number. For example: square of '5' is 5^2.

More Examples:

- $4^3 = 4 \times 4 \times 4 = 64$

  4^3 can be read as "4 to the power of 3" or "4 to the third power" or simply "4 cubed"

Note: The exponent or power of '3' of a number is also called **cube** of that number. For example: cube of '5' is 5^3.

- $10^9 = 10 \times 10 \times 10 \times 10 \times 10 \times 10 \times 10 \times 10 \times 10 = 1000000000$

  Thus, exponents make it easier to write and use lots of multiplications.

  In general, a to the power of n is denoted as –

$$a^n = \underbrace{a \times a \times \ldots \times a}_{n \text{ times}}$$

Negative Exponents :

A negative exponent or negative power of a number denotes how many times to divide 1 by that number. For example:

$$8^{-1} = 1 \div 8 = \frac{1}{8}$$

$$5^{-3} = 1 \div 5 \div 5 \div 5 = \frac{1}{5 \times 5 \times 5} = \frac{1}{5^3}$$

$$10^{-4} = 1 \div 10 \div 10 \div 10 \div 10 = \frac{1}{10 \times 10 \times 10 \times 10} = \frac{1}{10^4}$$

Properties:

1. For any number a, $a^0 = 1$
    Examples: $2^0 = 1$; $8^0 = 1$; $100^0 = 1$; $19876^0 = 1$

2. For any number a, $a^m \times a^n = a^{m+n}$
    Examples: $2^2 \times 2^3 = 2^{2+3} = 2^5 = 2 \times 2 \times 2 \times 2 \times 2 = 32$

3. For any number a, $(a^m)^n = a^{mn} = (a^n)^m = a^{nm}$

4. For any two numbers a and b, $\frac{a^m}{b^m} = \left(\frac{a}{b}\right)^m$

5. For any two numbers a and b:

$$\left(\frac{a}{b}\right)^m = \frac{a^m}{b^m}$$

$$\left(-\frac{a}{b}\right)^m = \frac{(-a)^m}{(+b)^m}$$

6. For any number a, $a^{-n} = \frac{1}{a^n}$

7. For any number a, $\frac{a^m}{a^n} = a^m \times a^{-n} = a^{m-n}$

8. For any number a, $(-a)^m = + a^m$ if m is even

9. Any power m of 1, $(1)^m = \underbrace{1 \times 1 \times \ldots \times 1} = 1$

10. Any power m of 0, $(0)^m = \underbrace{0 \times 0 \times \ldots \times 0}_{m \text{ times}} = 0$

11. For any two numbers a and b,

    if $a^m = b$ then, $a = b^{\frac{1}{m}}$

Note: Be careful about grouping:

To avoid confusion, use brackets ( ) in cases like this:

With ( ) :	$(-2)^2 = (-2) \times (-2) = 4$
Without ( ) :	$-2^2 = -(2^2) = - (2 \times 2) = -4$
With ( ) :	$(ab)^2 = ab \times ab$
Without ( ) :	$ab^2 = a \times (b)^2 = a \times b \times b$

4.7 Root

The root of a number 'a' is another number, which when multiplied by itself a given number of times, equals 'a'.

For example: the second root of 16 is 4, because 4 × 4 = 16

the fifth root of 32 is 2, because 2 × 2 × 2 × 2 × 2 = 32

The mth root of a number a is denoted as $\sqrt[m]{a}$. So, 5th root is denoted as $\sqrt[5]{a}$; 8th root is denoted as $\sqrt[8]{a}$.

Note:

i. The second root of a number is called **square root**. For any number a, square root of a is denoted as $\sqrt[2]{a}$ or simply $\sqrt{a}$

ii. The third root of a number is called **cube root**. For any number a, square root of a is denoted as $\sqrt[3]{a}$

Therefore, $\sqrt{a} \times \sqrt{a} = \left(\sqrt{a}\right)^2 = \left(a^2\right)^{\frac{1}{2}} = a^{\frac{1}{2} \times 2} = a$

The square root or second root when multiplied two times, equals the number.

$\sqrt[3]{a} \times \sqrt[3]{a} \times \sqrt[3]{a} = \left(\sqrt[3]{a}\right)^3 = \left(a^3\right)^{\frac{1}{3}} = a^{\frac{1}{3} \times 3} = a$

The cube root or third root when multiplied three times, equals the number.

Similarly, $\underbrace{\sqrt[m]{a} \times \sqrt[m]{a} \times \dots \times \sqrt[m]{a}}_{m \text{ times}} = \left(\sqrt[m]{a}\right)^m = \left(a^{\frac{1}{m}}\right)^m = a^{\frac{1}{m} \times m} = a$

The mth root when multiplied m times, equals the number.

Properties:

1. For any number a, the mth root of a, $\sqrt[m]{a} = a^{\frac{1}{m}}$

    Square root of $a = \sqrt[2]{a} = \sqrt{a} = a^{\frac{1}{2}}$

    Cube root of $a = \sqrt[3]{a} = a^{\frac{1}{3}}$

2. For any number a, the mth root of a^m , $\sqrt[m]{a^m} = a$

    $\sqrt[2]{a^2} = \sqrt{a^2} = a$

    or $\qquad \sqrt[2]{(\text{number})^2} = \sqrt{(\text{number})^2} = \text{number}$

    $\sqrt[3]{a^3} = a$

    or $\qquad \sqrt[3]{(\text{number})^3} = \text{number}$

Example: $\sqrt[5]{2^5} = (2^5)^{\frac{1}{5}} = 2^{5 \times \frac{1}{5}} = 2^1 = 2$

3. For any two numbers a and b,

$$\text{if } a^m = b \text{ then, } a = b^{\frac{1}{m}} = \sqrt[m]{a}$$

4.8 Mathematical Expression

A mathematical expression is a combination of numbers, symbols and operators (such as ×, +, etc.). For example: 1 + 2 + 3 is an expression.

Note: An expression does not contain the equal to (=) sign.

An expression may contain different operations. For example:

$2 \times 3 + 6 \div 3 \times 2^2$ is also an expression. It contains multiplication, addition, division and square operations.

In order to group certain parts of an expression together, brackets are used. **Brackets** are a pair of symbols used to enclose sections of a mathematical expression. There are four types of brackets:

——	Vinculum or bar or line brackets
( )	Small brackets or Round brackets or Parentheses
{ }	Middle brackets or Curly brackets or Braces
[ ]	Big brackets or Square brackets or Box brackets

For example: $5 \times [(3 + 2) \times (6 - 4)]$

4.9 Order of Operations

If an expression involves multiple additions, subtractions multiplications, divisions, etc. then we need to determine which part to calculate first. For example:

Calculate $2 + 3 \times 4$

If we calculate 2 + 3 first, we get –

$2 + 3 \times 4 = 5 \times 4 = 20$

If we calculate 3 × 4 first, we get –

$2 + 3 \times 4 = 2 + 12 = 14$

These are two different answers and only one of them is correct. To determine the order in which operations should be performed a rule called **BODMAS** is followed. According to BODMAS rule, the order in which operations are performed is:

B	Brackets first	
	( ) or { } or [ ]	
O	Of	
	(orders, i.e. Powers and Square Roots, Cube Roots, etc.)	
D	Division	
	÷	
M	Multiplication	
	×	
A	Addition	
	+	
S	Subtraction	
	–	

Thus, the order of operations, in an expression, can be listed as :–

- Solve the part in brackets first.
- Orders, i.e. powers and roots, before multiply, divide, add or subtract.
- Divide, before multiply, add or subtract.
- Multiply before add or subtract
- Add before subtract
- Finally subtract.

For example:

	$7 + (6 \times 5^2 + 3)$	Start inside **B**rackets and then solve **O**rders first
=	$7 + (6 \times 25 + 3)$	Then **M**ultiply
=	$7 + (150 + 3)$	Then **A**dd
=	$7 + (153)$	Brackets completed, final operation is add
=	$7 + 153$	(Answer)
=	160	

4.10 Divisibility

A number a is said to be divisible by another number b, if the remainder of the division a ÷ b is 0.

4.11 Even Numbers

The numbers divisible by '2' are called even numbers. For example – 0, 2,4 and 6 are even numbers.

4.12 Odd Numbers

The numbers not divisible by '2' are called odd numbers. For example – 1, 3, 5 and 7 are odd numbers.

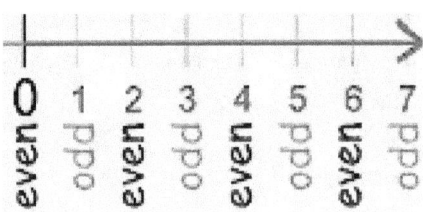

4.13 Fractions

Fractions are composed of a numerator and a denominator. The denominator tells us how many parts the numerator is divided into. For example, if you have one apple and wish to share it with someone, you need to cut it in half. Here, 1 is the numerator which is divided into 2 (denominator) parts, and as a result each person gets part of the whole. The number is a fraction.

Types of Fractions

1. **Proper Fraction :** A fraction whose numerator is less than its denominator is called a proper fraction. For example:

   $$\frac{3}{4}, \frac{5}{7}, \frac{6}{11}, \frac{14}{19}, \frac{20}{27}, \text{etc.}$$

2. **Improper Fraction :** A fraction whose numerator is greater than its denominator is called an improper fraction. For example:

   $$\frac{5}{4}, \frac{7}{5}, \frac{8}{3}, \frac{11}{6}, \frac{20}{17}, \text{etc.}$$

3. **Mixed Fraction :** A proper fraction combined with a whole number is called a mixed fraction. . For example:

   $$1\frac{2}{7} = 1 + \frac{2}{7} = \frac{7+2}{7} = \frac{9}{7} \text{ (improper fraction) ;}$$

   $$1\frac{3}{5} = 1 + \frac{3}{5} = \frac{5+3}{5} = \frac{8}{5} \text{ (improper fraction) ;}$$

   $$2\frac{3}{4} = 2 + \frac{3}{4} = \frac{4+3}{4} = \frac{7}{4} \text{ (improper fraction) ;}$$

   $$4\frac{5}{7} = 4 + \frac{5}{7} = \frac{7+5}{7} = \frac{12}{7} \text{ (improper fraction)}$$

4. **Unit Fraction :** A fraction with numerator equal to '1' is called a unit fraction. For example:

   $$\frac{1}{1}, \frac{1}{2}, \frac{1}{3}, \frac{1}{7}, \frac{1}{10}, \frac{1}{16}, \text{etc.}$$

5. **Like Fraction :** Fractions with same denominator are known as like fractions. For example:

$$\frac{2}{9}, \frac{3}{9}, \frac{4}{9}, \frac{5}{9}, \text{etc.}$$

6. **Unlike Fraction :** Fractions with different denominators are known as unlike fractions. For example:

$$\frac{1}{2}, \frac{1}{4}, \frac{3}{8}, \frac{5}{7}, \text{etc.}$$

4.14 Decimal Numbers

A decimal number contains a decimal point (**.**). A decimal number consists of two parts: a whole part and a fractional part.

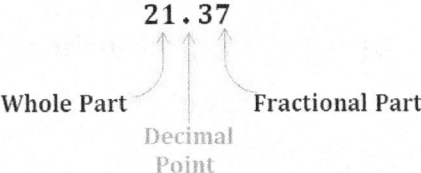

For example:

The part to the left of the decimal point forms the whole part, and the part to the right of the decimal point constitutes the fractional part. In the above example, 21 is the whole part and 37 is the fractional part.

A decimal number can be expressed as a sum of its whole and fractional parts as:

$$21.37 = 21 + 0.37$$

$$1.245 = 1 + 0.245$$

$$67.8 \ \ = 67 + 0.8$$

A whole number can be written as a decimal number with its fractional part as zero. For example:

$$5 = 5.0$$

$$108 = 108.0$$

Any preceding group of zeroes to the left of a whole number can be ignored. For example:

005 = 0000000005 = 5 = 5.0 = 00005.0

0000108 = 0108 = 108 = 108.0 = 000108.0

Any trailing zeros after the decimal point, i.e. in the fractional part can be ignored. For example:

0.5 = 0.500 = 00.50000 = 00000.500000

1.2 = 1.20 = 01.200 = 00001.200000 = 000000001.2000

2.0 = 2 = 002.000 = 00002.000000000

Note: The fractional part of a decimal number is read digit by digit. The abovementioned decimal number can be read as twenty one point three seven. Another example: the number 30.02445 can be read as thirty point zero two four four five.

Multiplication of a decimal number by powers of 10:

When a decimal number is multiplied with 10, the decimal point shifts one digit to the right. Similarly, when a decimal number is multiplied with 100, the decimal point shifts right by two digits, and so on. For example:

$$0.3125 \times 100 = 31.25$$

$$0.3125 \xrightarrow{\times 100} 31.25$$

$$0.322 \times 1000 = 322.0 = 322$$

$$0.322 \xrightarrow{\times 1000} 322$$

Fractional Representation of a Decimal Number:

A decimal number can be represented in terms of a fraction by following the steps mentioned below.

Step 1 : Write down decimal number divided by 1.

Step 2 : Multiply both top (numerator) and bottom (denominator) by 10 for every digit after the decimal point. For example: if there are two digits after the decimal point, then use 100; if there are three then use 1000, etc.)

Step 3 : Simplify the fraction.

For example: **Convert 0.75 into fraction**

Step 1 : $0.75 = \dfrac{0.75}{1}$

Step 2 : Since there are two digits after the decimal point, multiply both numerator and denominator with 100

$$\frac{0.75}{1} = \frac{0.75 \times 100}{1 \times 100} = \frac{75}{100}$$

Step 3 : Simplify

$$\frac{75}{100} = \frac{15}{20} = \frac{3}{4}$$

Therefore, $0.75 = \dfrac{3}{4}$

Another way of solving this is:

Step 1 : We know 1 = 1.0 = 1.000 = 1.00000. Since there are two digits after the decimal point in 0.75, add two trailing zeros to 1, i.e. 1.00. Now write 0.75 as 0.75 divided by 1.00.

$$0.75 = \frac{0.75}{1.00}$$

Step 2 : Since there are two digits after the decimal point, in both numerator and denominator, simply remove the decimal point in both.

$$\frac{0.75}{1.00} = \frac{075}{100} = \frac{75}{100}$$

Step 3 : Simplify

$$\overset{\div 5 \qquad \div 5}{\frac{75}{100} = \frac{15}{20} = \frac{3}{4}}_{\div 5 \qquad \div 5}$$

Therefore, $0.75 = \dfrac{3}{4}$

Recurring Decimal Number:

A recurring decimal number, also known as repeating decimal number, is a decimal number having digits repeating indefinitely. A recurring decimal number is represented by placing a bar (horizontal line) over the repeating digits. For example:

- $\dfrac{1}{3} = 0.3333... = 0.\overline{3}$

- $0.456456456... = 0.\overline{456}$

4.15 Rational Numbers

Any number that can be written as a fraction is a rational number. If "p" and "q" are two integers, then $\dfrac{p}{q}$ is a rational number. Here, q ≠ 0, since division by zero is undefined.

Thus, $\dfrac{1}{3}$ is a rational number. 3 is also a rational number since we can write it as $\dfrac{3}{1}$. Therefore, rational numbers include all integers and all fractions.

Note: Any number a has an implicit exponent of +1, an implicit denominator of +1 and an implicit sign of positive. It can be expressed as:

$$+1 \frac{a}{+1}^{+1}$$

Note: x^{-1} or $\frac{1}{x}$ is known as the reciprocal of x

4.16 Irrational Numbers

The numbers which are not rational, i.e. which cannot be expressed as a fraction, are known as irrational numbers. For example:

- The value of $\sqrt{2}$ equals 1.41421356237309504...

  It cannot be expressed in terms of a fraction. Hence, $\sqrt{2}$ is an irrational number
- The value of π (pi) equals 3.1415926535897932384...
- It cannot be written in terms of a fraction and thus, π is also an irrational number

4.17 Real Numbers

If we combine the set of rational numbers with the set of irrational numbers, we obtain a new set of numbers known as real numbers.

Thus, real numbers include:

- all the rational numbers, and
- all the irrational numbers

4.18 Factors

Factors are numbers we multiply together to get another number. A number can have many factors. For example:

1 × 12 = 12

2 × 6 = 12

3 × 4 = 12

Thus, **1, 2, 3, 4, 6** and **12** are the factors of 12. And also −1, −2, −3, −4, −6 and −12, because you get a positive number when you multiply two negatives, such as (−2)×(−6) = 12. Therefore, **1, 2, 3, 4, 6, 12, −1, −2, −3, −4, −6, −12** are the factors of 12.

Example −

Find all the factors of **18**.

Start with 1 : 1 × 18 = 18

Then go to 2 : 2 × 9 = 18

Then go to 3 : 3 × 6 = 18

Then go to 4 : 4 doesn't work since 4 × 4 = 16 and 4 × 5 = 20

Then go to 5 : 5 doesn't work either.

Then go to 6 : 6 × 3 = 18. Now factors are repeating.

Thus, the factors of 18 (after including negative numbers too) are **1, 2, 3, 6, 9, 18, –1, –2, –3, –6, –9, –18.**

4.19 Prime Numbers

A natural number greater than '1', having exactly two factors namely, '1' and itself, is known as a prime number. It does not have any other factors. For example: 2, 3, 5, 7, 11, 13, 17, 19, 23, 29, etc.

$2 = 2 \times 1 ; 3 = 3 \times 1; 5 = 5 \times 1$

In other words, a natural number greater than '1', divisible by only 1 and itself is prime. For example: 7 is divisible by only two numbers- 1 and 7.

The smallest prime number is 2. The only even prime number is 2, since all other even numbers will have one factor as 2.

4.20 Composite Numbers

A natural number greater than '1', having more than two factors is known as a composite number. For example: 4, 6, 8, 9, 12, 14, 15, 16, 18, etc.

$4 = 2 \times 2 \times 1 ; 6 = 2 \times 3 \times 1 ; 8 = 2 \times 2 \times 2 \times 2 \times 1$

In other words, a natural number greater than '1', divisible by any other number, except 1 and itself is composite. For example: 4 is divisible by 1, **2** and 4.

Note : '1' is neither prime nor composite.

4.21 Prime Factorization

The process of writing a number as a product of prime numbers is known as prime factorization. The prime numbers, which multiply together to form another number, are called the prime factors.

Example:

List the prime factors of 18.

Start with the smallest prime number, 2:

$18 \div 2 = 9$; We have the first prime factor as 2. But, 9 is not prime. So we need to break it further into product of primes.

We again divide 9 by 2 (smallest prime). 9 is not divisible by 2.

So, **we move to the next prime number**, i.e. **3**: $9 \div 3 = 3$. 3 is a prime number, so no need to break it further.

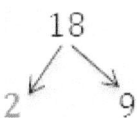

$\therefore 18 = 2 \times 3 \times 3$

The prime factors of 18 are **2, 3** and **3.**

4.22 Highest Common Factor (HCF)

When we find all the factors of two or more numbers, and some factors are the same ("common"), then the largest of those common factors is the Highest Common Factor (HCF). It is also called greatest common factor or greatest common divisor (GCD).

For example:

Find the HCF of 12 and 16

Method 1: Factorization

Factors of 12 = ①, ②, ③, 4 , 6, 12

Factors of 16 = ①, ②, ④, 8, 16

1, 2 and 4 are common factors

4 is the greatest or highest

The HCF of 12 and 16 is 4

Method 2: Prime Factorization

12 = ① × ② × ② × 3

16 = ① × ② × ② × 2 × 2

1, 2 and 2 are common prime factors.

HCF is the product of the common prime factors.

HCF of 12 and 16 = 1 × 2 × 2 = 4

Method 3: Division method

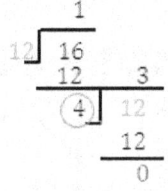

Step 1 : Divide the larger number by smaller one.

Divisor = 12; Dividend = 16

Step 2 : Next, the divisor is considered as dividend and the remainder as new divisor.

Step 3 : Divide

Step 4 : Repeat steps 2 and 3 until the remainder becomes 0.

The divisor, in the end, is the HCF.

HCF of 12 and 16 is 4.

Note: To find the HCF of three numbers by division method, first find the HCF of any two numbers. Then, find the HCF of the third number and HCF obtained in the previous step, to get the result. Similarly, HCF of four or more numbers can be calculated by considering them in pairs.

4.23 Twin Prime Pair

Two prime numbers with difference equal to '2' are called twin primes of one another. For example: (3,5), 3 and 5 are twin primes since 5 – 3 = 2 ; (5,7), 5 and 7 are twin primes since 7 – 5 = 2

4.24 Co-prime Numbers

Two integers are said to be co-prime, or relatively prime or mutually prime, if they have only one common factor, i.e. '1'. For example: 18 and 25 are co-prime numbers.

$18 = 2 \times 3 \times 3 \times \mathbf{1}$

$25 = 5 \times 5 \times \mathbf{1}$

4.25 Multiple

The result of multiplying a number by an integer (except 0), is called a multiple of that number. For example: multiples of 3 are 3, 6, 9, 12, ... 12 is a multiple of 3, since 12 = 3×4 (integer). Note that 0 is not a multiple of 3. Also note that –3, –6, –9, –12, ... are also multiples of 3. –6 is a multiple of 3, since –6 = 3×-2 (integer).

Note: 0 is not a multiple of any other number except 0 itself.

4.26 Least Common Multiple (LCM)

The smallest positive integer, which is a multiple of two or more numbers, is called the Least Common Multiple (LCM). It is also known as lowest common multiple.

For example –

Find the LCM of 4 and 10

Method 1: Listing all the Multiples

Multiples of 4 = 4, 8, 12, 16, ⑳, 24, 28, 32, 36, ㊵, ...

Multiples of 10 = 10, ⑳, 30, ㊵, ...

20 and 40 are two common multiples

20 is the least.

The LCM of 4 and 10 is 20

Method 2: Prime Factorization

$4 = 1 \times 2 \times 2 = 1 \times 2^2 \times 5^0$

$10 = 1 \times 2 \times 5 = 1 \times 2^1 \times 5^1$

The product of the sets of primes with the highest exponent value, in among all prime factorizations, is the LCM.

LCM of 4 and 10 = $2^2 \times 5^1 = 4 \times 5 = 20$

Method 3: using Common Prime Factors grid

Follow the same steps as prime factorization, but consider two or more numbers (whose LCM has to be calculated) at the same time. Thus, instead of finding the prime factors of a number, we find the common prime factors of two or more numbers.

LCM is the product of the common prime factors of the given numbers.

LCM of 4 and 10 = $2 \times 2 \times 5 = 20$

Common Prime Factors		
2	4	10
2	2	5
5	1	5
	1	1

Method 4: HCF method

For any two positive numbers a and b

$$\textbf{product of two numbers} = \boldsymbol{a} \times \boldsymbol{b} = \textbf{HCF} \times \textbf{LCM}$$

Therefore, $\text{LCM} = \dfrac{\text{Product of two numbers}}{\text{HCF}}$

We can use the above relation to find the LCM of two numbers.

HCF of 4 and 10 = 2

$\text{LCM of 4 and 10} = \dfrac{4 \times 10}{\text{HCF of 4 and 10}} = \dfrac{40}{2} = 20$

Note: For any three positive numbers a, b and c,

$$\textbf{product of three numbers} = \boldsymbol{a} \times \boldsymbol{b} \neq \textbf{HCF} \times \textbf{LCM}$$

Thus, to find the LCM of three numbers by HCF method, first find the HCF of two numbers, and then use that HCF to find the LCM of those two numbers. Then find the LCM of the LCM of those two numbers and the third number. Similarly, the LCM of four or more numbers can be calculated.

Roman Numerals

Ancient Romans used aspecial method for writing numbers known as the roman numerical system.

5.1 The Roman Symbols

Roman numerals are based on the following seven symbols:

Roman Symbol	I	V	X	L	C	D	M
Hindi-Arabic Number	1	5	10	50	100	500	1000

Note that there is no symbol or digit for zero.

5.2 Forming Roman Numerals

Rule 1: The roman digits I, X and C are repeated up to three times in succession to form the numbers.

We know the value of I is 1, X is 10 and C is 100. The values of I, X and C are added as:

I = 1   X = 10

II = 1 + 1 = 2   XX = 10 + 10 = 20

III = 1 + 1 + 1 = 3   XXX = 10 + 10 +10 = 30

C = 100

CC = 100 + 100 = 200

CCC = 100 + 100 + 100 = 300

Note that the digits V, L and D are not repeated.

Rule 2: When the symbol I, X or C appears **after** a symbol, it is **added.**

VI = 5 + 1 = 6   LX = 50 + 10 = 60

XI = 10 + 1 = 11   DC = 500 + 100 = 600

Rule 3: When the symbol I, X or C appears **before** a symbol, it is **subtracted**.

IV = 5 – 1 = 4	XC = 100 – 10 = 90
IX = 10 – 1 = 9	CD = 500 – 100 = 400
XL = 50 – 10 = 40	CM = 1000 – 100 = 900

5.3 Example

Hindi Arabic Numeral System to Roman Numeral System

The approach is to break the number into thousands, hundreds, tens and ones, and then write roman numerals for each in turn.

Convert 1949 to Roman Numerals.

1949 = 1000 + 900 + 40 + 9

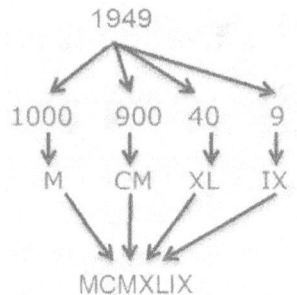

1949 is broken into 1000, 900, 40 and 9, and then each conversion is done as:

- M is the roman symbol for 1000, 1000 = M
- Conversion for 500 is D, 600 is DC (500 + 100), 700 is DCC and 800 is DCCC. C can be repeated a maximum of three times, thus 900 = 1000 – 100 = CM
- Conversion for 10 is X, 20 is XX and 30 is XXX. X can be repeated up to three times, thus 40 = 50 – 10 = XL
- Conversion for 5 is V, 6 is VI (5 + 1), 7 is VII and 8 is VIII. I can be repeated a maximum of three times, thus 9 = 10 – 1 = IX

The individual conversions are combined to get the Roman numeral for 1949 as MCMXLIX.

Roman Numeral System to Hindi Arabic Numeral System

The approach is:

- Find the symbol having the largest value
- If it is the first symbol, check if it is repeated in succession
- If the it is at second position, convert the combined value of first two symbols into Arabic system
- Repeat this process for the leftover Roman numeral.

Convert CDXXXIX to Hindi Arabic Numeral System

Largest Symbol is D. It is the second symbol, so CD is converted to Arabic system as CD = 500 – 100 = **400**. Leftover Roman numeral is XXXIX.

Largest symbol in leftover Roman numeral is X. X is the first symbol, and is repeated in succession three times, XXX = 10 + 10 + 10 = **30**. Leftover Roman numeral = IX.

Largest symbol in leftover Roman numeral is X. It occurs at second position, so IX is converted to Arabic system as IX = 10 – 1 = **9**.

The combined number in Hindi Arabic system is 400 + 30 + 9 = **439**.

5.4 Basic Formations

I	II	III	IV	V	VI	VII	VIII	IX
1	2	3	4	5	6	7	8	9

X	XX	XXX	XL	L	LX	LXX	LXXX	XC
10	20	30	40	50	60	70	80	90

C	CC	CCC	CD	D	DC	DCC	DCCC	CM
100	200	300	400	500	600	700	800	900

5.5 Numbers greater than 1000

Numbers greater than 1000 are formed by placing a dash (or horizontal line) over the symbols. If a horizontal line is drawn over the symbols, then the value of the numerals becomes 1000 times.

$V = 5$ but $\overline{V} = 5000$,

$XV = 15$ but $\overline{XV} = 15000$,

$CLV = 155$ but $\overline{CLV} = 155000$

Algebra

Algebra (or alzebra) is a branch of mathematics, which deals with the study of mathematical symbols and the rules for manipulating these symbols.

6.1 The Concept of a Variable

Unlike arithmetic, which involves only known numerical values, algebra is based on the notion of unknown values, called variables. A variable may be denoted by any letter in English alphabet (a to z and A to Z). It is usually denoted as x. A variable can take any numerical value. For example: x – 4 = 5;

Here, x is a variable, i.e. unknown. It can take any numerical value, say 1, 2, 3, etc. But, for the above relation to be true, x should take the value 9, since 9 – 4 = 5

The numerical values in an algebraic expression are also known as constants, since they have a fixed value, which does not change. For example, the number 9 always represents the same value.

6.2 Equation

An equation, in algebra, can be thought of as a scale, such as the one shown below. But, instead of balancing weights, it balances numbers and variables.

Several symbols, such as operation symbols (+, –, ×, ÷) , brackets, etc. are used to combine the variables and numbers together.

For example:

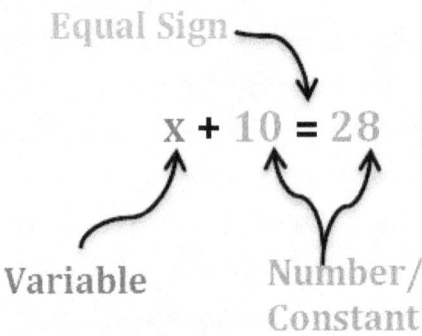

The equal sign (=) is analogous to the balancing point of the scale, i.e. the portion of the equation on the left hand side (LHS) of the equal sign (=) should be balanced against the portion on the right hand side (RHS) of the equal sign (=).

6.2.1 Solving an Equation

Solving an equation means to evaluate the value of all the unknowns, i.e. variables, in that equation. To solve the above equation for x, we should subtract the LHS with 10, since x + 10 – 10 will reduce to x. But, if we subtract 10 from LHS, we need to subtract 10 from RHS as well, since the LHS and RHS should remain balanced.

x + 10 – 10 = 28 – 10

x + 0 = 18

x = 18

Thus, the value of x is obtained as 18.

Note: As long as the same operation, such as addition, subtraction, multiplication, etc. is carried out on both the sides of =, i.e. LHS and RHS, the equation will remain balanced.

Let us consider a more complex equation:

4x – 10 = x + 5

Subtracting x from both the sides, we get –

4x – 10 – x = x + 5 – x

(4x – x) – 10 = (x – x) + 5

3x – 10 = 0 + 5

3x – 10 = 5

Adding 10 to both the sides, we get –

3x – 10 + 10 = 5 + 10

3x = 15

3 × x = 15

Now there are two approaches for solving:

Approach 1:

Dividing both sides by 3, we get –

$$\frac{\cancel{3} \times x}{\cancel{3}} = \frac{\cancel{15}^{5}}{\cancel{3}_{1}}$$

x = 5

Approach 2:

Multiplying both sides with , we get –

$$3 \times x \times \frac{1}{3} = 15 \times \frac{1}{3}$$

$$x \times \cancel{3} \times \frac{1}{\cancel{3}} = \cancel{15}^{5} \times \frac{1}{\cancel{3}}$$

x = 5

Thus, the value of x is 5.

Tips for solving an equation:

1. Any term (variable or constant) in an equation, is moved across the '=' according to some rules. **When variables or numbers are moved from one side of '=' to the other**:

   - addition (+) **becomes** subtraction (−)
   - subtraction (−) **becomes** addition (+)
   - multiplication (×) **becomes** division (÷)
   - division (÷) **becomes** multiplication (×)

   For example: In the equation,

   4x + 3 = 7

   If we move ' + 3 ' to the RHS of '=', it becomes '− 3'

   4x = 7 − 3

2. **Bring the unknown (or variable) on one side of '='** and the numerical values on the other side of '='.

   For example: when solving an equation such as:

   4x + 3 = 2x + 8

   We will first bring 2x on the LHS of '=' so that both the unknown terms, i.e. 4x and 2x are on the same side of '='.

   4x − 2x = 8 − 3

3. **Coefficient of a variable** is the number by which that variable is multiplied in an equation. For example:

   - In 4x + 5 = 3, the coefficient of x is 4
   - In 0.854x − 5 = 7, the coefficient of x is 0.854
   - In 6 − 5x = 10, the coefficient of x is −5
   - In 5 + x = 4, the coefficient of x is 1

   An equation may have more than one variable. For example:

   4x + 3y = 10 (Equation with two vaiables: x and y)

   x + y + z = 100 (Equation with three variables: x,y and z)

Note:

If we have just one unknown variable, we can solve for its value with only one equation. If we have two variables, we need at least two equations to determine both values. For example:

x + y = 10 is an equation with two variables x and y. Using just this equation we cannot determine the values of x and y. x and y can be any combination of values with sum equal to 10, such as 5 and 5; 6 and 4; 3 and 7; and so on. Thus we need another equation. If the other equation is:

x − y = 0

x = y (−y on LHS becomes +y on RHS)

Putting x = y in equation x + y = 10

y + y = 10

2y = 10

y = 10/2 (×2 on LHS becomes ÷2 on RHS)

y = 5

Since x = y; x = 5

Similarly, **in order to solve for '*n*' variables, we need a minimum of '*n*' equations.**

6.2.2 Substitution:

Substitution means replacing a variable with its value, i.e. a number. For example: if we have x – 4, and we know that x = 5; then we can substitute x with 5 and obtain x – 4 = 5 – 4 = 1

If x = 5, y = 3, then calculate x + 2y

x + 2y = 5 + 2 × 3   (Substituting x with 5 and y with 3)

= 5 + 6

= 11

Note: When substituting variables with negative numbers, always place them in brackets. For example:

If x = −3; Calculate $x^2 + 2x$

$x^2 + 2x = (-3)^2 + 2(-3)$

= 9 + (−6)

= 9 − 6 = 3

since $(-3)^2 \neq -3^2$; it is important to put brackets when substituting with negative numbers.

6.3 Polynomials

The word polynomial is derived from *poly* meaning 'many', and *nomial* meaning 'term'. Thus, polynomial means 'many terms'. Polynomials are sums of "variables and exponents" expressions. Each part of a polynomial that is being added, is called a "term."

Polynomial terms can have:

- **constants**: such as 4, $\frac{1}{2}$, −10

- **variables:** such as x, y, z. The variables can have coefficients such as 4x, −3y, $\frac{1}{2}Z$
- **exponents:** variables raised to only whole number exponents are allowed, such as 0, 1, 2, and so on. For example: x^2, $5x^3$, etc.

  Note that, square roots of variables (i.e. exponent of), fractional powers, and negative powers do not form polynomial

For example:

$\frac{1}{x^2} = x^2$ is not a polynomial term, since the variable x has negative exponent.

$\sqrt{x}$ is not a polynomial term, since the variable x is raised to a non whole number power, i.e. $\frac{1}{2}$

7 is a polynomial term, since it is a constant.

An example of a typical polynomial is:

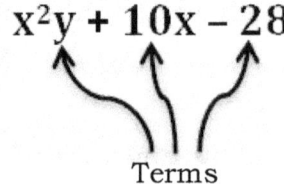

Each term of a polynomial is separated by a + or – sign.

Note the exponents on the terms. In the first term x has an exponent of 2 and y has an exponent of 1; in the second term x has an exponent of 1; and the last term doesn't have any variable. **Polynomials are usually written in this manner, wherein the terms appear in decreasing value of exponents of variables.**

There are special names for polynomials with one, two and three terms:

Monomial	Polynomial with 1 term Example: $5x^2y$
Binomial	Polynomial with 2 terms Example: $5y - 4$
Trinomial	Polynomial with 3 terms Example: $3y^3 + 9y - 4$

Note that: A Polynomial can have as many terms as needed, but not an infinite number of terms.

6.3.1 Degree of a Polynomial

The degree of a polynomial with **only one variable** is the **largest exponent** of that variable in that polynomial. For example:

$5y - 4$; degree = 1

$3y^3 + 9y - 4$; degree = 3

If a polynomial has **more than one variable**, we need to inspect each term. The **degree of each term** is calculated by **adding the exponents** of each variable in it; and the **largest such degree** among all the terms is the degree of polynomial. For example:

Consider $5x^2y + 2y^2 + 6$

Degree of $5x^2y$ = 2 (exponent of x) + 1 (exponent of y) = 3

Degree of $2y^2$ = 2 (exponent of y)

Degree of 6 = 0 (since no variable)

The largest degree is 3; thus, degree of the polynomial is 3.

Note: A polynomial of degree 0 is a constant.

6.4 Linear Equation

An equation with degree equal to 1 is called a linear equation. For example:

- $y = 4x + 3$
- $y - 2 = 6 - 2x$
- $5z = 10$
- $x - 2y + 4z - 7 = 0$

$x^2 - 2x = y$ is not a linear equation since it has a degree of 2.

Note: A linear equation has only one solution, i.e. only one value of x for which the equation is satisfied.

6.5 Quadratic Equation

An equation of degree 2 is known as a quadratic equation. The standard notation of a quadratic equation is:

$$ax^2 + bx + c = 0$$

where, a, b and c are constants such that $a \neq 0$; and x is the variable. Examples of quadratic equations:

- $2x^2 + x - 4 = 0$; a = 2, b = 1 and c = −4
- $x^2 - 5 = 0$; a = 1, b = 0 and c = −5
- **$x^2 = 5x - 2$ can be written as $x^2 - 5x + 2 = 0$; a = 1, b = −5 and c = 2**

If $a = 0$, the equation $bx + c = 0$ is not a quadratic equation, since its degree is 1 (and not 2).

Note: A quadratic equation with b = c = 0, but a ≠ 0 is known as a **pure quadratic equation**. For example: $4x^2 = 0$.

6.5.1 Roots of a Quadratic Equation

The solutions of a quadratic equation, i.e. the value (or values) of unknown variable which makes it equal to zero, are known as the **roots or zeroes** of a quadratic equation. **A quadratic equation can have at most two roots or solutions.**

The roots of a quadratic equation, $ax^2 + bx + c = 0$ are:

$$x = \frac{-b + \sqrt{b^2 - 4ac}}{2a}, \text{ and}$$

$$x = \frac{-b - \sqrt{b^2 - 4ac}}{2a}$$

Together the roots of a quadratic equation can be written as:

$$x = \frac{-b \pm \sqrt{b^2 - 4ac}}{2a}$$

The abovementioned formula for roots of a quadratic equation is called the **quadratic formula**. The expression **b^2-4ac** is known as the **discriminant**, since it can discriminate among different cases.

Case 1 : When **b^2-4ac > 0** : The quadratic equation has **two distinct real roots** given by the quadratic formula.

Case 2 : When **b^2-4ac = 0** : the quadratic equation has only **one real root** (the two roots are **equal**).

Case 3 : When **< 0** : the quadratic equation has **no real roots** (imaginary roots).

For example:

Find the roots of the quadratic equation $5x^2 + 6x + 1 = 0$

Comparing the given equation with standard notation, we get: a = 5, b = 6, c = 1

Discriminant, $b^2 - 4ac = (6)^2 - 4 \times 5 \times 1 = 36 - 20 = 16$

Since $b^2 - 4ac > 0$, the quadratic equation has two real roots, given by:

$$x = \frac{-b \pm \sqrt{b^2 - 4ac}}{2a}$$

$$= \frac{-6 \pm \sqrt{16}}{2 \times 5} = \frac{-6 \pm 4}{10} = \frac{-6 + 4}{10} \text{ or } \frac{-6 - 4}{10}$$

$$= \frac{-2}{10} \text{ or } \frac{-10}{10} = -0.2 \text{ or } -1$$

Formation of a Quadratic Equation when two roots are given:

If **a** and **b** are two roots of a quadratic equation, then the required quadratic equation can be formed as:

$$x^2 - (a + b)x + ab = 0$$

or

$$x^2 - \textbf{(sum of roots)}x + \textbf{(product of roots)} = 0$$

Note: Similarly, an equation of degree 3 is called a cubic equation. For example: $x^3 - 4x^2 + 10$ is a cubic equation with degree 3.

6.6 Algebraic Identities

$(a + b)^2 = a^2 + b^2 + 2ab$

$(a + b)^2 = (a + b)^1(a + b)^1$

$$\begin{array}{r} a + b \\ a + b \\ \hline a^2 + ab \\ + ab + b^2 \\ \hline a^2 + 2ab + b^2 \end{array}$$

$(a - b)^2 = a^2 + b^2 - 2ab$

$(a - b)^2 = (a - b)^1 (a - b)^1$

$$
\begin{array}{r}
+\, a - b \\
+\, a - b \\
\hline
a^2 - ab \\
- ab + b^2 \\
\hline
a^2 - 2ab + b^2
\end{array}
$$

$(a - b)(a + b) = a^2 - b^2$

$$
\begin{array}{r}
+\, a - b \\
+\, a + b \\
\hline
a^2 - ab \\
+ ab - b^2 \\
\hline
a^2 \quad\ - b^2
\end{array}
$$

$(a + b)^3 = a^3 + b^3 + 3ab(a + b)$

$(a + b)^3 = (a + b)^2(a + b) = (a^2 + b^2 + 2ab)(a + b)$

$$
\begin{array}{r}
a^2 + b^2 + 2ab \\
a + b \\
\hline
a^3 + a^2b \\
+ ab^2 + b^3 \\
+ 2a^2b + 2ab^2 \\
\hline
a^3 + a^2b + ab^2 + b^3 + 2a^2b + 2ab^2
\end{array}
$$

$= a^3 + b^3 + 3a^2b + 3ab^2$

$= a^3 + b^3 + 3ab(a + b)$

$(a - b)^3 = a^3 - b^3 - 3ab(a - b)$

$(a - b)^3 = (a - b)^2(a - b) = (a^2 + b^2 - 2ab)(a - b)$

$$
\begin{array}{r}
+\, a^2 + b^2 - 2ab \\
+ a - b \\
\hline
a^3 - a^2b \\
+ ab^2 - b^3 \\
- 2a^2b + 2ab^2 \\
\hline
a^3 - a^2b + ab^2 - b^3 - 2a^2b + 2ab^2
\end{array}
$$

$= a^3 - b^3 - 3a^2b + 3ab^2$

$= a^3 - b^3 - 3ab(a - b)$

$(a + b)(a^2 - ab + b^2) = a^3 + b^3$

$$a + b$$
$$a^2 - ab + b^2$$

$$\overline{}$$

$$a^3 - a^2b + ab^2$$
$$+ a^2b - ab^2 + b^3$$

$$\overline{}$$

$$a^3 \qquad\qquad + b^3$$

$(a - b)(a^2 + ab + b^2) = a^3 - b^3$

$$a - b$$
$$a^2 + ab + b^2$$

$$\overline{}$$

$$a^3 + a^2b + ab^2$$
$$- a^2b - ab^2 - b^3$$

$$\overline{}$$

$$a^3 \qquad\qquad - b^3$$

Mensuration

7.1 Line

A straight path passing through points A and B, indefinitely extending in two opposite directions is called a **straight line**. It is denoted by $\overleftrightarrow{AB}$

It can also be denoted by small letters, such as m.

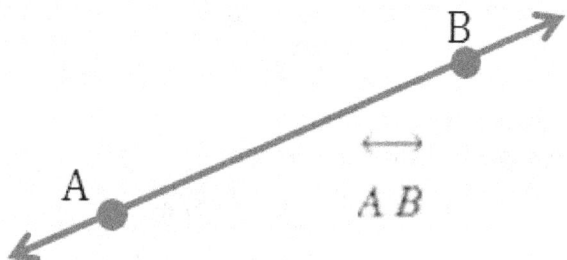

The shortest distance between two points A and B is called a **line segment**. It is denoted by $\overline{AB}$ or $\overline{BA}$.

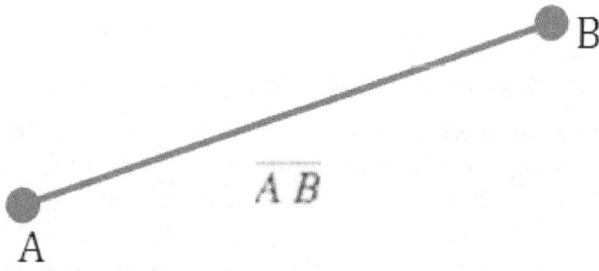

A straight line having one end point at A and extending indefinitely beyond another end point B is called a **ray** AB. It is denoted a by $\overrightarrow{AB}$

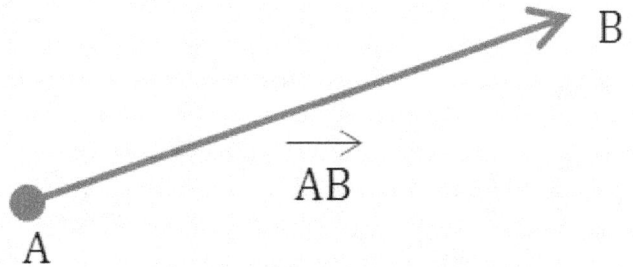

7.2 Parallel Lines

If two lines never meet, they are known as parallel lines.

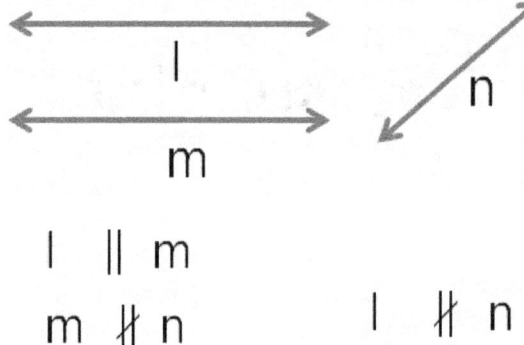

$$l \parallel m$$
$$m \nparallel n$$
$$l \nparallel n$$

7.3 Angle

If two straight lines have a common point then they form an angle. The two straight lines are called the sides of the angle, and the common point is called the vertex of the angle.

It is denoted by ∠. The angle shown in figure can be denoted as ∠ABC or ∠CBA or ∠B.

AB and BC are the two rays forming the angle

∠ABC, with B as the common endpoint. B is called the vertex of ∠ABC.

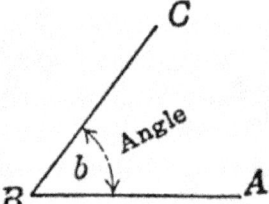

7.3.1 Types of Angles

There are five different types of angles.

1. **Acute Angle :** An angle whose measure is less than 90°.

2. **Right Angle :** An angle whose measure is equal to 90°.

3. **Obtuse Angle :** An ngle whose measure is greater than 90° but less than 180°.

4. **Straight Angle :** An angle whose measure is equal to 180°. Straight angle looks like a straight line.

5. **Reflex Angle :** An angle whose measure is greater than 180° but less than 360°.

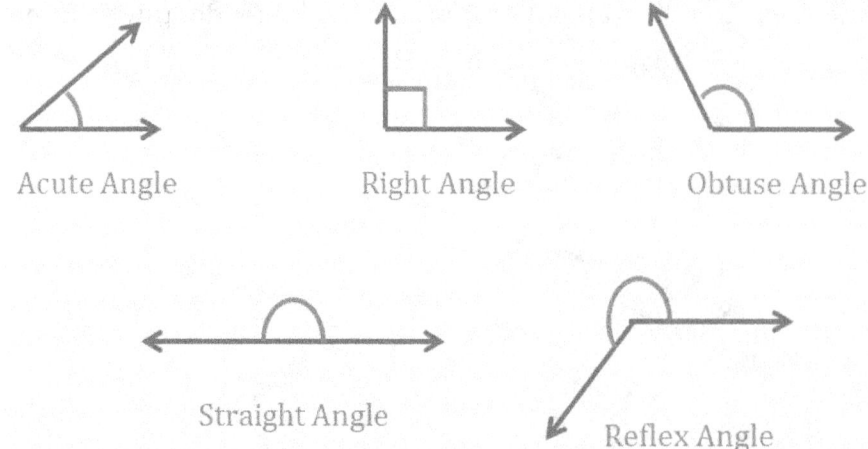

Acute Angle   Right Angle   Obtuse Angle

Straight Angle   Reflex Angle

7.3.2 Angles and their Laws

Complementary Angles :

If the sum of two angles is 90°, they are known as complementary angles. For example:

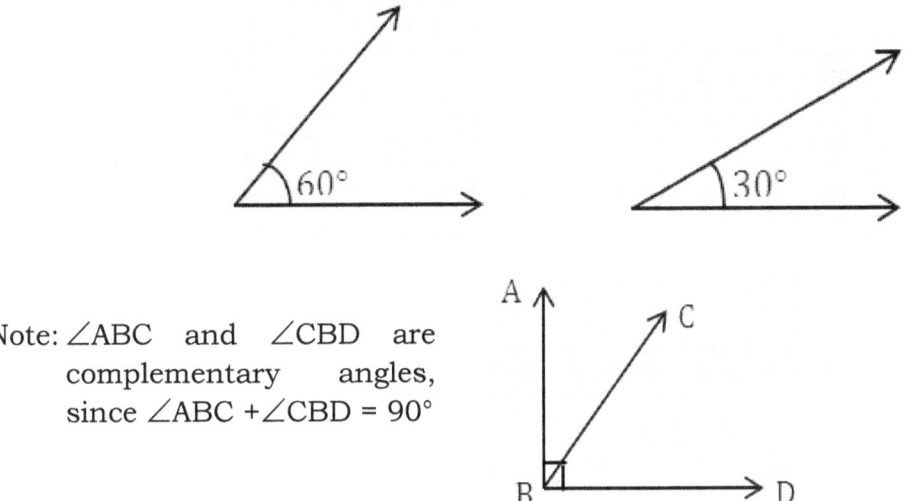

Note: ∠ABC and ∠CBD are complementary angles, since ∠ABC +∠CBD = 90°

Supplementary Angles

If the sum of two angles is 180°, they are known as supplementary angles. For example:

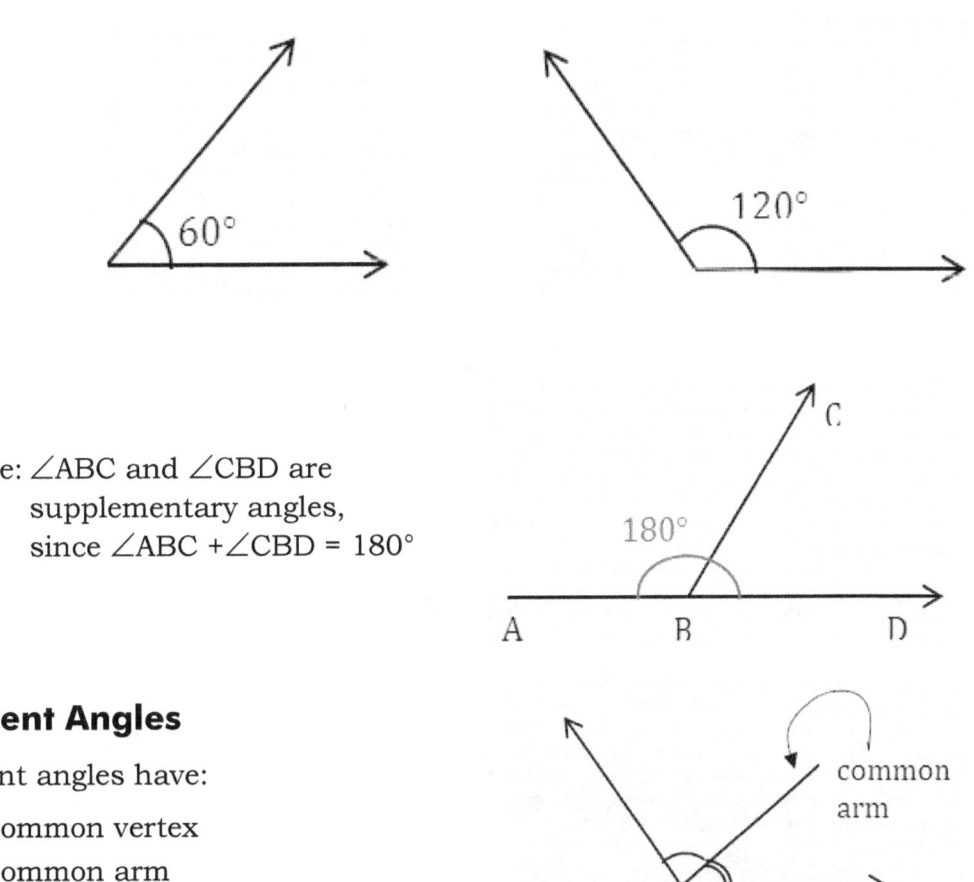

Note: ∠ABC and ∠CBD are supplementary angles, since ∠ABC +∠CBD = 180°

Adjacent Angles

Adjacent angles have:

i.   a common vertex
ii. a common arm
iii. a non-common arm on either side

Vertically Opposite Angles

When two lines intersect, the vertically opposite angles so formed are equal.

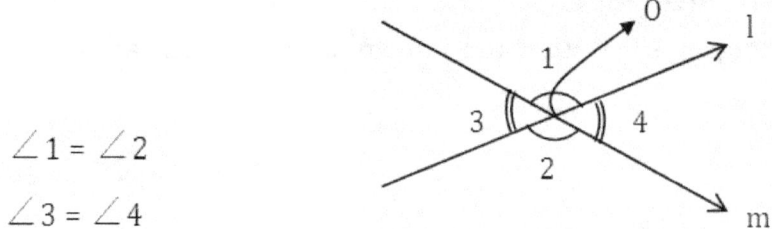

$$\angle 1 = \angle 2$$

$$\angle 3 = \angle 4$$

Point of intersection: It two lines, l and m, intersect and if they have a common point (O), then O is the point of intersection.

Transversal

A line 'p' intersecting two or more lines at distinct (distant) points is known as a transversal.

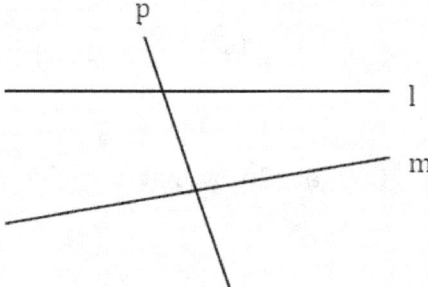

The lines, l and m, may be parallel as well.

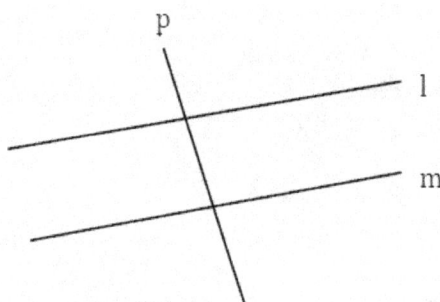

Transversal Angles :

The following groups of angles can be identified from the figure below:

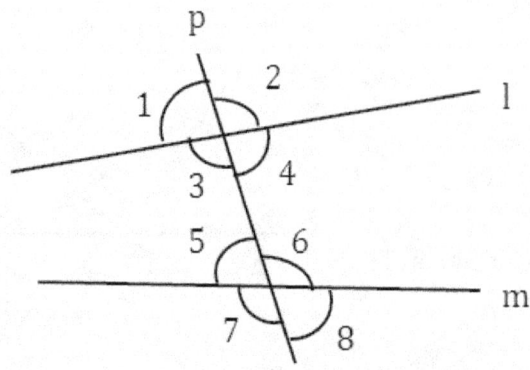

i.   Interior Angles: $\angle 3$, $\angle 4$, $\angle 5$ and $\angle 6$

ii. Exterior Angles: $\angle 1$, $\angle 2$, $\angle 7$ and $\angle 8$

iii. Pair of Corresponding Angles: $\angle 1$ and $\angle 5$; $\angle 2$ and $\angle 6$; $\angle 3$ and $\angle 7$; $\angle 4$ and $\angle 8$

iv. Pair of Alternate Interior Angles: $\angle 3$ and $\angle 6$; $\angle 4$ and $\angle 5$

v.   Pair of Alternate Exterior Angles: $\angle 1$ and $\angle 8$; $\angle 2$ and $\angle 7$

vi. Pair of Interior Angles on the Same Side of Transversal: $\angle 3$ and $\angle 5$; $\angle 4$ and $\angle 6$

If the lines l and m are parallel, the following laws hold:

i.   Pair of corresponding angles are equal. $\angle 1 = \angle 5$; $\angle 2 = \angle 6$; $\angle 3 = \angle 7$; $\angle 4 = \angle 8$

ii. Pair of alternate interior angles are equal. $\angle 3 = \angle 6$; $\angle 4 = \angle 5$

iii. Pair of alternate exterior angles are equal. $\angle 1 = \angle 8$; $\angle 2 = \angle 7$

iv. Pair of interior angles on the same side of transversal are supplementary. $\angle 3 + \angle 5 = 180°$; $\angle 4 + \angle 6 = 180°$

7.4 Curve

A and B are points at the two ends of the line AB. The path A to B through either of the points C or D is a curve.

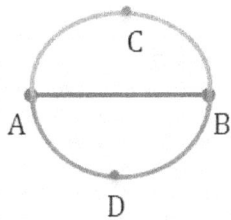

ACB and ADB are two curves. A and B are the two endpoints of the curves ACB and ADB.

Curves are of two types :

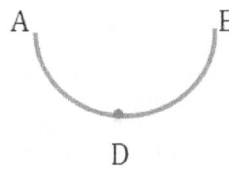

1. **Closed Curve** : If the two endpoints of a curve meet, it is called a closed curve. ACBDA is a closed curve.

2. **Open Curve** : If the two endpoints of a curve do not meet, it is called an open curve. ACB and ADB are open curves.

7.5 Region

The interior of a closed curve together with its boundary is known as a region. The outer portion, excluding the region, is called exterior.

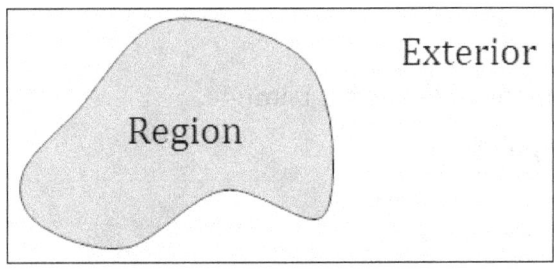

7.6 Polygon

A closed curve made from straight line segments is called a polygon. A few examples of polygons are -

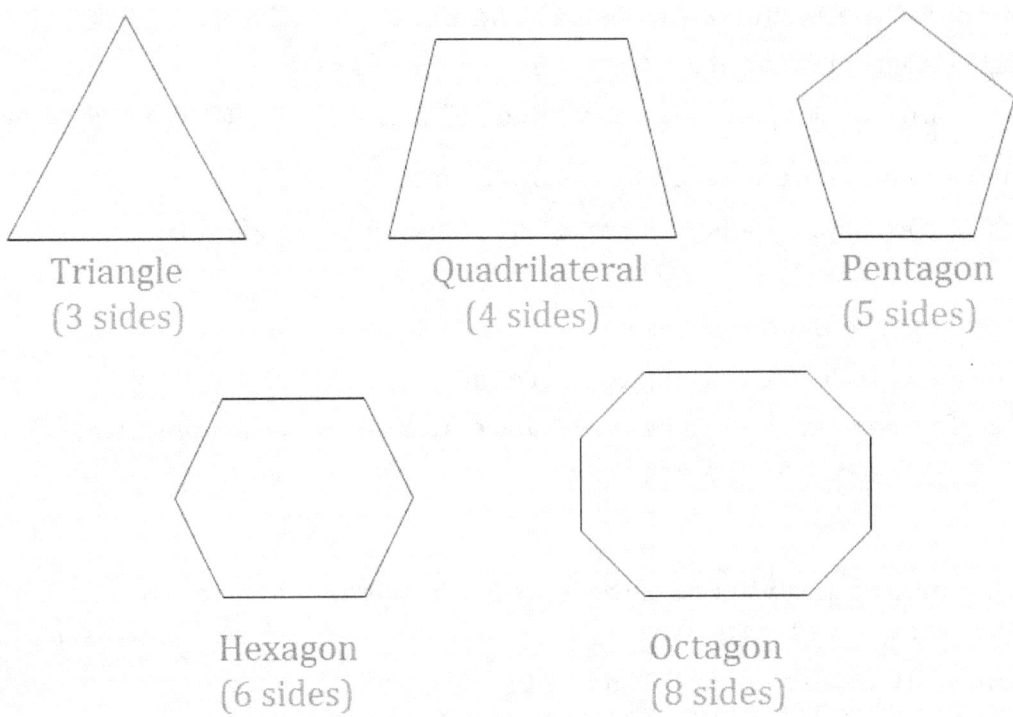

Triangle
(3 sides)

Quadrilateral
(4 sides)

Pentagon
(5 sides)

Hexagon
(6 sides)

Octagon
(8 sides)

The following figure is not a polygon due to curved nature at the top.

7.7 Triangle

A figure bounded by three sides is called a triangle.

A triangle is a three-sided polygon.

ΔABC, in figure, has –

- three vertices A, B and C
- three sides AB, BC and CA
- three angles ∠ABC, ∠BCA and ∠CAB (We can also denote
- ∠ABC, ∠BCA and ∠CAB as ∠B, ∠C and ∠A respectively.)

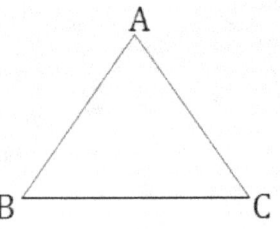

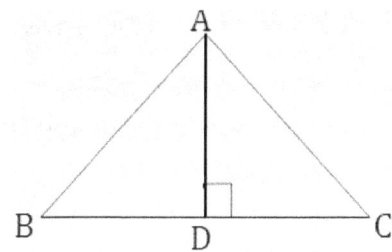

Area of a $\Delta = \dfrac{1}{2} \times$ base $\times$ corresponding altitude

$\qquad = \dfrac{1}{2} \times$ BC $\times$ AD

Perimeter of a Δ = Sum of three sides = AB + BC + CA

Sum of angles of a $\Delta = 180°$

or $\angle$ABC + $\angle$BCA + $\angle$CAB = 180°

or $\angle$A + $\angle$B + $\angle$C = 180°

Heron's Formula for area of a triangle: The area of a triangle with side lengths a, b and c is given by : –

$$A = \sqrt{s\,(s-a)(s-b)(s-c)}$$

where, s represents the semi-perimeter of the triangle, given as : –

$$s = \frac{a+b+c}{2}$$

7.7.1 Types of Triangles based on Sides

1. **Equilateral Triangle:**

   An equilateral triangle has all three sides (or all three angles) equal.
   - AB = BC = CA
   - $\angle$A = $\angle$B = $\angle$C = 60°

   a is the length of a side of equilateral triangle.

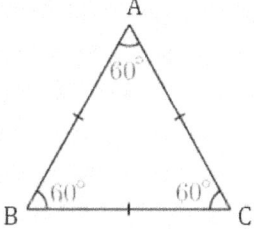

   Area of an equilateral triangle $= \dfrac{\sqrt{3}}{4} a^2$

2. **Isosceles Triangle:**

   An isosceles triangle has exactly two sides (or two angles) equal. The remaining third side (or third angle) is different.
   - Two sides are equal
     AB = AC
   - Two angles are equal
     $\angle$B = $\angle$C

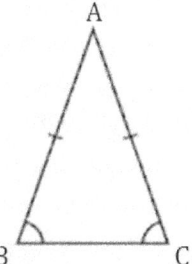

3. **Scalene Triangle:**

   A triangle with no equal sides or no equal angles is called a scalene triangle.
   - No equal sides
     AB ≠ BC ≠ CA
   - No equal angles
     $\angle$A ≠ $\angle$B ≠ $\angle$C

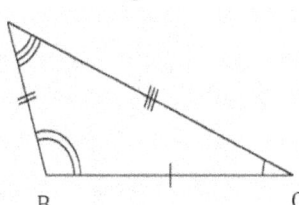

7.7.2 Types of Triangles based on Angles

1. Right Angled Triangle :

A right angled triangle has one of its angles equal to 90°.

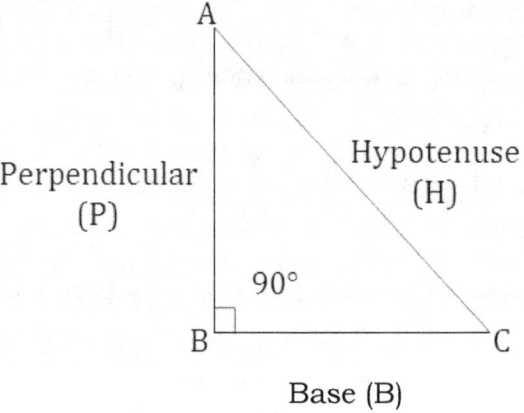

$$H^2 = P^2 + B^2$$
$$\text{or} \quad H = \sqrt{P^2 + B^2}$$

$\angle A + \angle B + \angle C = 180°$
If $\angle B = 90°$, then
$\angle A + 90° + \angle C = 180°$
$\angle A + \angle C = 180° - 90° = 90°$

Area $= \dfrac{1}{2} \times B \times P$

1(a) Isosceles Right Angled Triangle :

An isosceles right angled triangle has one right angle (90°) and the other two angles/two sides equal.

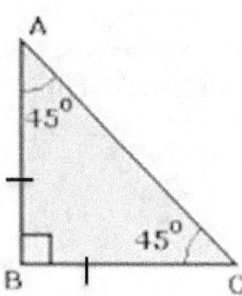

2. Obtuse Triangle :

An obtuse angled triangle has one of its angles greater than (>) 90°. It has an obtuse angle in the triangle.

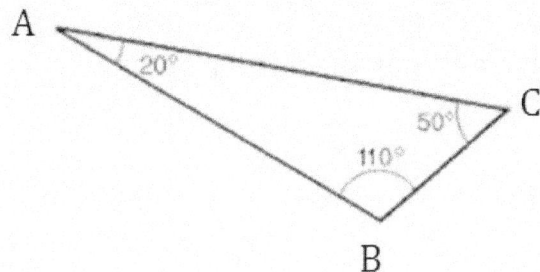

3. Acute Triangle :

An acute angled triangle has all three angles less than (<) 90°. It has all three angles as acute angles.

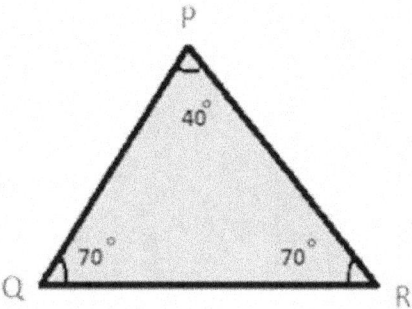

7.8 Quadrilateral

A quadrilateral is a four-sided polygon. The four sides may be equal or unequal.

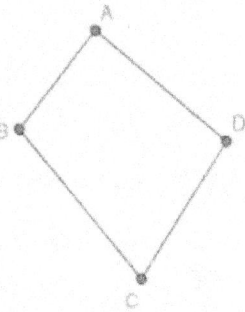

    Perimeter = AB + BC + CD + DA

    Sum of angles = $\angle$A + $\angle$B + $\angle$C + $\angle$D

                = 360°

7.9 Parallelogram

A parallelogram is a quadrilateral in which opposite sides are parallel and equal in length.

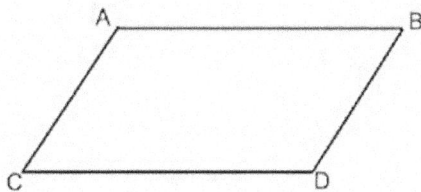

AB = CD

AC = BD

AB || CD

AC || BD

Two angles ($\angle$B and $\angle$C) are less than 90°.

Two angles ($\angle$A and $\angle$D) are greater than 90°.

7.10 Rhombus

A rhombus is a parallelogram with all sides equal.

7.11 Square

A square has four sides of equal length and four corners, each of which is a right angle (90°).

AB = BC = CD = DA = a

Area = side × side

$\quad$ = (side)²

$\quad$ = **a²**

Perimeter = AB + BC + CD + DA

Since AB = BC = CD = DA,

Perimeter = 4 × one side = **4 × a**

Sum of angles = ∠A + ∠B + ∠C + ∠D

Since ∠A = ∠B = ∠C = ∠D = 90°,

Sum of angles = 4 × one angle

$\quad\quad$ = 4 × 90°

$\quad\quad$ = **360°**

Note that a square is a parallelogram with all sides and all angles equal. A rhombus with all angles equal to 90° becomes a square.

7.12 Rectangle

A rectangle has four sides, of which opposite sides are equal, and four angles, each equal to 90°.

AB = CD = L (Length)

BC = AD = B (Breadth)

Area = L × B

Perimeter = AB + BC + CD + AD

$\quad\quad$ = L + B + L + B

$\quad\quad$ = 2L + 2B

$\quad\quad$ = **2(L + B)**

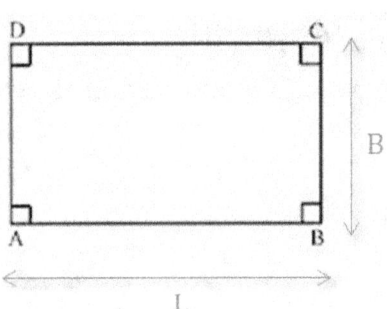

Sum of angles = ∠A + ∠B + ∠C + ∠D = 4 × 90° = **360°**

Note that a rectangle is a special type of parallelogram in which each of the corner angles is 90°. So, all rectangles are parallelograms but all parallelograms are not rectangles.

Note that a square is a type of rectangle with L = B = a.

7.13 Circle

A circle is a shape having all points at the same distance from its center. The figure shown below is a circle.

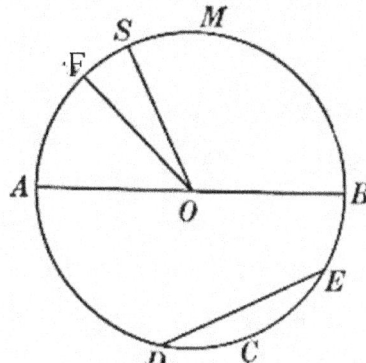

- The point **O** is called the center of the circle.
- All the points **A, B, C, D, E, F, M** and **S** are at same distance from the center **O**, known as the radius (**R**) of circle. **Radius (R) of circle = OA**
- **DE** is known as a chord. A chord is a line segment joining any two points on the circle.
- **AB** is called the diameter of the circle. It is twice the radius (R).
  Diameter (D) of circle = AB = 2 × R
- **Revolution :** One complete turn of a point on the circle either clockwise or anti-clockwise is known as one revolution.
- 1 complete revolution = 360° = 2π radians (**where π = 3.14**)
- $\frac{1}{2}$ of a revolution = 180° (two right angles) = π radians
- $\frac{1}{4}$ of a revolution = 90° (one right angle) = $\frac{\pi}{2}$ radians

$$\textbf{Area of a circle} = \pi R^2 = \pi\left[\frac{D}{2}\right]^2 = \frac{\pi D^2}{4}$$

Circumference of a circle = 1 complete revolution × radius

$$= 2\pi \times \textbf{R} = \pi\,(2\textbf{R}) = \pi\textbf{D}$$

7.14 Cube

A cube (dice) is a solid figure bounded by six identical square plane surfaces. Each square plane surface is called the face of the cube. The point at which three edges (sides) of a cube meet is known vertex.

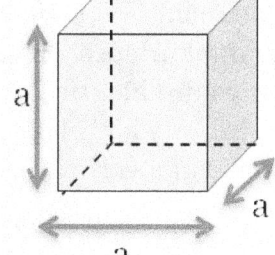

- Number of faces = 6
- Number of vertices = 8
- Number of edges (sides) = 12

a is the length of one edge (side).

Area of one face (square surface) of a cube = a × a = a²

Lateral or Curved Surface Area of a cube

= sum of areas of square faces except top and bottom

= a² + a² + a² + a²

= **4** × **a²**

= **4** × **(edge)²**

Total Surface Area of a cube (S)

= sum of areas of all the square faces

= a² + a² + a² + a² + a² + a²

= **6** × **a²**

= **6** × **(edge)²**

Perimeter of one square surface of a cube = 4a

Total perimeter of a cube = 12a

Volume of a cube (V) = area of one surface × height

= a² × a = **a³**

= **(edge)³**

Volume, V = (edge)³ = a³

or a³ = V

$$\boxed{\begin{array}{l} \text{or}\quad a = V^{\frac{1}{3}} \\ \text{or}\quad \text{edge} = (\text{Volume})^{\frac{1}{3}} \end{array}}$$

Surface Area (S) = 6 × (edge)² = 6a²

or 6a² = S

or $a^2 = \dfrac{S}{6}$

or $a = \dfrac{S^{\frac{1}{2}}}{6} = \sqrt{\dfrac{S}{6}}$

or edge = $\sqrt{\dfrac{\text{Surface Area}}{6}}$

7.15 Cuboid

A cuboid is a solid figure bounded by six rectangular plane surfaces. In a cuboid, the opposite rectangular faces are identical (equal in length and breadth). The point at which three edges (sides) of a cuboid meet is known as vertex.

- Number of faces = 6

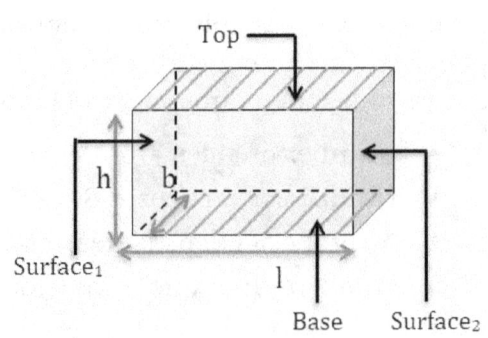

- Number of vertices = 8
- Number of edges (sides) = 12

l is the length, **b** is breadth and **h** is height of cuboid.

Since both base and top of cuboid are rectangular,

Area of base = l × b

Area of top = l × b

Area of surface$_1$ = b × h

Area of surface opposite to surface$_1$ = b × h

Area of surface$_2$ = l × h

Area of surface opposite to surface$_2$ = l × h

Lateral or Curved Surface Area of cuboid

= sum of areas of rectangular faces except top and bottom

= l×h + l×h + b×h + b×h

= 2× (l×h + b×h)

= 2h (l + b)

Total Surface Area of cuboid (S)

= sum of areas of all the rectangular faces

= l×b + l×b + b×h + b×h + l×h + l×h

= 2 × lb + 2 × lh + 2 × bh

= 2 × (lb + bh + lh)

If l = b = h = a (cube)

Total surface area (S) = 2 × (a×a + a×a + a×a)

= 2 × (a^2 + a^2 + a^2)

= 2 × 3a^2

= 6 × a^2

Volume of a cuboid (V) = area of base × height

= l × b × h

If l = b = h = a (cube)

Volume (V) = a × a × a = a^3

Volume, V = l × b × h

or $l = \dfrac{V}{b \times h}$

or $\text{length} = \dfrac{\text{volume}}{\text{breadth} \times \text{height}}$

or $b = \dfrac{V}{l \times h}$

or $length = \dfrac{volume}{length \times height}$

or $h = \dfrac{V}{b \times l}$

or $height = \dfrac{volume}{breadth \times length}$

7.16 Right Circular Cylinder

A right circular cylinder is a solid figure having two circular plane surfaces, one at its base and another at its top. The two circular plane surfaces are identical in all respects (equal radius). It has a curved surface in the middle.

- Number of faces = 2
- Number of vertices = 0
- Number of edges = 0

r is the radius of circular plane surfaces of the cylinder and **h** is the height of the cylinder.

Area of one circular surface = πr^2

Total area of both circular surfaces (at top and bottom) = $\pi r^2 + \pi r^2$

$= 2\pi r^2$

Lateral or Curved Surface Area = Area of middle portion

= circumference of circular surface × height

$= 2\pi r \times h$

Total Surface Area of cylinder (S)

= area of both circular surfaces + lateral surface area

$= 2\pi r^2 + 2\pi rh$

$= 2\pi r(h + r)$

Volume of cylinder (V) = area of circular surface × height

$= \pi r^2 \times h$

7.17 Hollow Right Circular Cylinder

If a right circular cylinder of smaller radius is cut off from the center of a right circular cylinder of larger radius, the resultant solid figure (excluding the cut off portion) is termed as a hollow cylinder. A few examples of hollow cylinders are pipes, tubes, band rings, etc.

h is the height of hollow cylinder, **r** is the radius of the inner cylinder and **R** is the radius of the outer cylinder.

Lateral Surface Area or Curved Surface Area

= external surface area + internal surface area

$$= 2\pi Rh + 2\pi rh = \mathbf{2\pi(R+r)h}$$

Area of top ring = area of outer circle – area of inner circle = $\pi R^2 - \pi r^2$

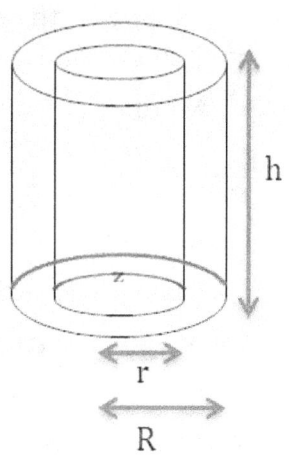

$$= \pi(R^2 - r^2)$$

Area of bottom ring = area of top ring

Total are of two base rings = $2 \times \pi(R^2 - r^2)$

Total Surface Area = curved surface area + area of two base rings

$$= 2\pi(R+r)h + 2\pi(R^2 - r^2)$$

Volume = volume of outer cylinder – volume of inner cylinder

$$= \pi R^2 h - \pi r^2 h = \mathbf{\pi(R^2 - r^2)h}$$

7.18 Right Circular Cone

A right circular cone is a solid figure formed by a circular base and a vertex at the top. The middle portion of a cone is curved.

- Number of faces = 1
- Number of vertices = 1
- Number of edges = 0

r is the radius of circular base of cone, **h** is the height of cone and **l** is the slant height or lateral height of cone.

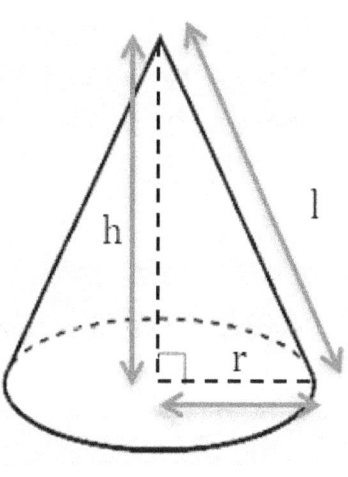

In a right angled triangle,

$$\text{Hypotenuse}^2 = \text{Perpendicular}^2 + \text{Base}^2$$

or $$\text{Hypotenuse} = \sqrt{\text{Perpendicular}^2 + \text{Base}^2}$$

∴ Lateral height, $l = \sqrt{h^2 + r^2}$

Lateral or Curved Surface Area = πrl

Total Surface Area (S) = curved surface area + area of circular base

$$= \pi rl + \pi r^2$$

$$= \mathbf{\pi r(l + r)}$$

Volume of a cone (S) = $\dfrac{1}{3}\pi r^2 h$

7.19 Sphere

A sphere is a solid figure formed by all the points that are at the same distance from the center point. Like a circle, a sphere has a radius and a diameter.

The shape of the earth is like a large sphere -- it has radius of about 4000 miles. A tennis ball is a sphere with a radius of about 2.5 inches.

- Number of faces = 0
- Number of vertices = 0
- Number of edges = 0

r is the radius of sphere.

Lateral or Curved Surface area = 4πr²

Total Surface Area (S) = Curved Surface Area = 4πr²

Volume of sphere (V) = $\dfrac{4}{3}\pi r^3$

7.20 Hemisphere

Hemisphere is the exact half of a sphere. When a plane cuts across a sphere at its center it forms two equal hemispheres.

- Number of faces = 1
- Number of vertices = 0
- Number of edges = 0

r is the radius of hemisphere.

Lateral or Curved Surface area = 2πr²

Total Surface Area (S) = curved surface area + area of circular top

$$= 2\pi r^2 + \pi r^2$$

$$= 3\pi r^2$$

Volume of hemisphere (V) $= \dfrac{1}{2} \times$ volume of sphere

$$= \dfrac{1}{2} \times \dfrac{4}{3}\pi r^3$$

$$= \dfrac{2}{3}\pi r^3$$

7.21 Pyramid

A pyramid a solid figure with the base as a polygon and the sides as triangles, which meet at the top. The top vertex at which the triangular sides meet is called apex.

The figure shows a square pyramid (having square base). The base can be any polygon such as triangle, rectangle, pentagon and so on.

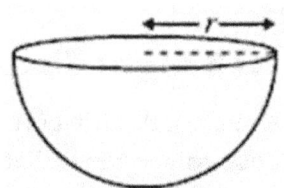

- Number of faces = at least 4
- Number of vertices = at least 4
- Number of edges = at least 6

7.21.1 Triangular Pyramid

A triangular pyramid has a triangle as its base and three triangles forming its sides.

- Number of faces = 4
- Number of vertices = 4
- Number of edges = 6

A **tetrahedron** is a triangular pyramid having identical equilateral trian its faces.

a is the side length of a tetrahedron.

Perimeter of triangular base = a + a + a = 3a

Lateral or Curved Surface area of a tetrahedron

= sum of the areas of three identical equilateral triangles

$$= 3 \times \frac{\sqrt{3}}{4} a^2$$

Total Surface Area (S) of a tetrahedron

$$= \text{area of base} + \text{curved surface area}$$
$$= \frac{\sqrt{3}}{4} a^2 + \frac{3\sqrt{3}}{4} a^2$$
$$= \frac{4\sqrt{3}}{4} a^2 = \sqrt{3}\, a^2$$

Volume (V) of a tetrahedron $= \frac{1}{3} \times \text{area of base} \times \text{height}$

$$= \frac{1}{3} \times \frac{\sqrt{3}}{4} a^2 \times \sqrt{\frac{2}{3}}\, a$$
$$= \frac{a^3}{6\sqrt{2}}$$

7.21.2 Square Pyramid

A square pyramid has a square as its base and four triangles forming its sides.

- Number of faces = 5
- Number of vertices = 5
- Number of edges = 8

a is the side length of square base, **h** is the height of pyramid and **l** is the lateral height of pyramid.

Perimeter of square base = 4a

In a right angled triangle,

$$\text{Hypotenuse}^2 = \text{Perpendicular}^2 + \text{Base}^2$$
$$\text{or} \quad \text{Hypotenuse} = \sqrt{\text{Perpendicular}^2 + \text{Base}^2}$$

$$\therefore \quad \text{Lateral height, } l = \sqrt{h^2 + \left[\frac{a}{2}\right]^2}$$

$$= \sqrt{h^2 + \frac{a^2}{4}}$$

Lateral or Curved Surface area

$$= \frac{1}{2} \times \text{Perimeter of base} \times \text{lateral height}$$

$$= \frac{1}{2} \times 4a \times \sqrt{h^2 + \frac{a^2}{4}}$$

$$= 2a\sqrt{h^2 + \frac{a^2}{4}} = a\sqrt{4h^2 + a^2}$$

Total Surface Area (S) = area of base + curved surface area

$$= a^2 + 2a\sqrt{h^2 + \frac{a^2}{4}}$$
$$= a^2 + a\sqrt{4h^2 + a^2}$$

7.22 Frustum of a Cone

If a plane cuts off a right circular cone parallel to its base, then the portion between the cutting plane and the base of the cone is known as a frustum of the cone. A frustum of a cone has two circular plane surfaces, one at top and another at bottom, of different radii.

- Number of faces = 2
- Number of vertices = 0
- Number of edges = 0

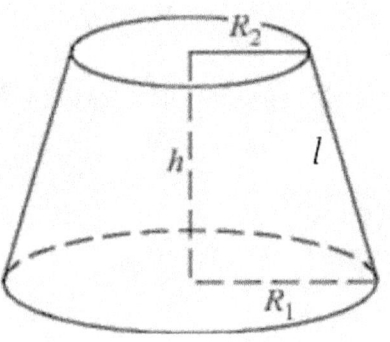

h is the height of frustum, R_1 is the radius of lower base, R_2 is the radius of upper base and *l* is the slant height or lateral height of frustum.

Note that $R_1 > R_2$.

Slant height, $l = \sqrt{h^2 + (R_1 - R_2)^2}$

Lateral Surface Area or Curved Surface Area = $\pi(R_1 + R_2)l$

Total Surface Area = $\pi[(R_1 + R_2)l + R_1{}^2 + R_2{}^2]$

Volume (V) = $\frac{\pi}{3}\left[R_1{}^2 + R_1 R_2 + R_2{}^2\right]h$

Conversion Units

8.1 Units of Time

1 Hour	=	60 Minutes
1 Minute	=	60 Seconds
A.M.	=	Ante Meridian
P.M.	=	Post Meridian
1 Day	=	24 Hours

The year divisible by '4' is known as a leap year. A leap year has 29 days in February.

1 Month =
28 (February, Non Leap Year)
29 (February, Leap Year)
30
31

1 Year = 12 Months =
365 Days (Non Leap Year)
366 Days (Leap Year)

8.2 Units of Currency

1 Rupee (₹) = 100 Paise

Coin = 10 Paisa, 25 Paisa, 50 Paisa, 100 Paisa

8.3 Units of Distance

1 km (kilometer)	=	1000 m (meter)
1 m	=	100 cm (centimeter)
1 cm	=	10 mm (millimeter)
1 cm	=	$\dfrac{1}{100}$ m
1 Mile	=	1760 Yard
1 Yard	=	3 Feet
1 Foot	=	12 Inches

8.4 Units of Weight

1 l (litre)	=	1000 ml (milli-litre)
1 l	=	1000 cc (centimeter cube)
1 Ton	=	10 Quintals
1kg (kilogram)	=	1000 g (gram)

8.5 Units of Angle

$$1° \text{ (degree)} = \frac{\pi}{180} \text{ radians } (\pi = 3.14)$$

$$1 \text{ radian} = \frac{180°}{\pi} \text{ (degree)}$$

$1°$ (degree) $= 60'$ (minutes)

$1'$ (minute) $= 60''$ (seconds)

$1°$ (degree) $= 3600''$ (seconds)

Note that DMS form means Degree Minutes Second form.

8.6 Indian vs International Number Representation

Indian	Inter-national	Numbers (Indian)	Numbers (Inter-national)	Powers of 10
One	One	1	1	10^0
Ten	Ten	10	10	10^1
Hundred	Hundred	100	100	10^2
Thousand	Thousand	1,000	1,000	10^3
Ten Thousand	Ten Thousand	10,000	10,000	10^4
Lakh	Hundred Thousand	1,00,000	100,000	10^5
Ten Lakh	Million	10,00,000	1,000,000	10^6
Crore	Ten Million	1,00,00,000	10,000,000	10^7
Ten Crore	Hundred Million	10,00,00,000	100,000,000	10^8
Arab	Thousand Million	1,00,00,00,000	1,000,000,000	10^9
Ten Arab	Ten Thousand Million	10,00,00,00,000	10,000,000,000	10^{10}
Kharab	Hundred Thousand Million	1,00,00,00,00,000	100,000,000,000	10^{11}
Ten Kharab	Billion	10,00,00,00,00,000	1,000,000,000,000	10^{12}

| Sankh | Ten Billion | 1,00,00,00,00,00,000 | 10,000,000,000,000 | 10^{13} |
| Maha Sankh | Hundred Billion | 10,00,00,00,00,00,000 | 100,000,000,000,000 | 10^{14} |

8.7 Metric Prefixes and Symbols

PREFIX	SYMBOL	MULTIPLIER
exa-	E	$1\ 000\ 000\ 000\ 000\ 000\ 000 = 10^{18}$
peta-	P	$1\ 000\ 000\ 000\ 000\ 000 = 10^{15}$
tera-	T	$1\ 000\ 000\ 000\ 000 = 10^{12}$
giga-	G	$1\ 000\ 000\ 000 = 10^{9}$
mega-	M	$1\ 000\ 000 = 10^{6}$
kilo-	k	$1\ 000 = 10^{3}$
hecto-	h	$100 = 10^{2}$
deca-	D (or da)	$10 = 10^{1}$
deci-	d	$0.1 = 10^{-1}$
centi-	c	$0.01 = 10^{-2}$
milli-	m	$0.001 = 10^{-3}$
micro-	µ	$0.000\ 001 = 10^{-6}$
nano-	n	$0.000\ 000\ 001 = 10^{-9}$
pico-	p	$0.000\ 000\ 000\ 001 = 10^{-12}$
femto-	f	$0.000\ 000\ 000\ 000\ 001 = 10^{-15}$
atto-	a	$0.000\ 000\ 000\ 000\ 000\ 001 = 10^{-18}$

Kilo (k) = 1000 = 10^3

1 kilogram (kg)   = 1000 grams (g)   = $10^3\,g$

1 kilometer (km) = 1000 meters (m) = $10^3\,m$

1 kilojoule (kJ)   = 1000 joules (J)   = $10^3\,J$

1 kilohertz (kHz) = 1000 hertz (Hz)   = $10^3\,Hz$

$$1\,g = 10^{-3}\,kg = \frac{1}{1000}\,kg = \frac{1}{10^3}\,kg$$

$$1\,m = 10^{-3}\,km = \frac{1}{1000}\,km = \frac{1}{10^3}\,km$$

$$1\,J = 10^{-3}\,kJ = \frac{1}{1000}\,kJ = \frac{1}{10^3}\,kJ$$

$$1\,Hz = 10^{-3}\,kHz = \frac{1}{1000}\,kHz = \frac{1}{10^3}\,kHz$$

Mega (M) = 1000000 = 10⁶ = Million

1 megagram (Mg)   = 1000000 grams (g)   = 10^6 g

1 megameter (Mm)   = 1000000 meters (m) = 10^6 m

1 megajoule (MJ)   = 1000000 joules (J) = 10^6 J

1 megahertz (MHz)   = 1000000 hertz (Hz) = 10^6 Hz

1 mega kilogram (Mkg) = 1000000 kg   = 10^6 kg

1 g   = 10^{-6} Mg   = $\dfrac{1}{1000000}$ Mg   = $\dfrac{1}{10^6}$ Mg

1 m   = 10^{-6} Mm   = $\dfrac{1}{1000000}$ Mm   = $\dfrac{1}{10^6}$ Mm

1 J   = 10^{-6} MJ   = $\dfrac{1}{1000000}$ MJ   = $\dfrac{1}{10^6}$ MJ

1 Hz = 10^{-6} MHz = $\dfrac{1}{1000000}$ MHz = $\dfrac{1}{10^6}$ MHz

1 kg = 10^{-6} Mkg = $\dfrac{1}{1000000}$ Mkg = $\dfrac{1}{10^6}$ Mkg

Deca (D or da) = 10

1 decagram (Dg)   = 10 grams (g)

1 decameter (Dm) = 10 meters (m)

1 decajoule (DJ) = 10 joules (J)

1 decahertz (DHz) = 10 hertz (Hz)

1 g   = 10^{-1} Dg   = $\dfrac{1}{10}$ Dg

1 m   = 10^{-1} dDm = $\dfrac{1}{10}$ Dm

1 J   = 10^{-1} DJ   = $\dfrac{1}{10}$ DJ

1 Hz = 10^{-1} DHz = $\dfrac{1}{10}$ DHz

Deci (d) = 10⁻¹

$$1 \text{ decigram (dg)} \quad = \frac{1}{10} \text{ g} \quad = 10^{-1} \text{ grams (g)}$$

$$1 \text{ decimeter (dm)} \quad = \frac{1}{10} \text{ m} \quad = 10^{-1} \text{ meters (m)}$$

$$1 \text{ decijoule (dJ)} \quad = \frac{1}{10} \text{ J} \quad = 10^{-1} \text{ joules (J)}$$

$$1 \text{ decihertz (dHz)} \quad = \frac{1}{10} \text{ Hz} \quad = 10^{-1} \text{ hertz (Hz)}$$

1 g = 10 dg

1 m = 10 dm

1 J = 10 dJ

1 Hz = 10 dHz

Milli (m) = 10⁻³

$$1 \text{ milligram (mg)} \quad = \frac{1}{1000} \text{ g} \quad = 10^{-3} \text{ grams (g)}$$

$$1 \text{ millimeter (mm)} \quad = \frac{1}{1000} \text{ m} \quad = 10^{-3} \text{ meters (m)}$$

$$1 \text{ millijoule (mJ)} \quad = \frac{1}{1000} \text{ J} \quad = 10^{-3} \text{ joules (J)}$$

$$1 \text{ millihertz (mHz)} \quad = \frac{1}{1000} \text{ Hz} \quad = 10^{-3} \text{ hertz (Hz)}$$

1 g = 1000 mg

1 m = 1000 mm

1 J = 1000 mJ

1 Hz = 1000 mHz

Micro (μ) = 10^{-6}

$$1 \text{ microgram (μg)} = \frac{1}{1000000} \text{ g} = 10^{-6} \text{ grams (g)}$$

$$1 \text{ micrometer (μm)} = \frac{1}{1000000} \text{ m} = 10^{-6} \text{ meters (m)}$$

$$1 \text{ microjoule (μJ)} = \frac{1}{1000000} \text{ J} = 10^{-6} \text{ joules (J)}$$

$$1 \text{ microhertz (μHz)} = \frac{1}{1000000} \text{ Hz} = 10^{-6} \text{ hertz (Hz)}$$

1 g = 1000000 μg = 10^6 μg

1 m = 1000000 μm = 10^6 μm

1 J = 1000000 μJ = 10^6 μJ

1 Hz = 1000000 μHz = 10^6 μHz

Trigonometry

The term trigonometry is derived from the Greek words *trigōnon* meaning "triangle" and *metron* meaning "measure." Trigonometry refers to a branch of mathematics, which deals with relationships between the sides and angles of triangles.

9.1 Right Angled Triangle

As discussed earlier, a right angled triangle has one of its angles equal to 90°. Consider the following right angled triangle, Δ ABC.

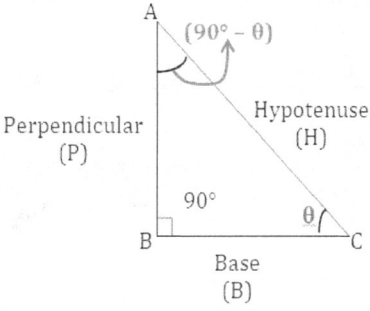

Let us consider an angle θ (read as theta), as shown in the figure. As we know, sum of the three angles of a triangle equals 180°.

$\angle A + \angle B + \angle C = 180°$

Since $\angle B = 90°$, we get –

$\angle A + 90° + \angle C = 180°$

$\angle A + \angle C = 180° - 90° = 90°$

Substituting $\angle C$ as θ, we get –

$\angle A + θ = 90°$

$\angle A = (90° - θ)$

The **hypotenuse** (H) of a right triangle is always the side opposite to right angle. It is the **longest side** in a right triangle.

The other two sides are called the perpendicular and base. These sides are labeled in relation to the angle being considered, i.e. θ.

The side opposite to the given angle θ is called the **perpendicular** (P). The non-hypotenuse side adjacent to the angle θ is called the **base** (B).

Note: If we consider a different angle, say (90°−θ), the perpendicular and base will change, but the hypotenuse will remain unchanged.

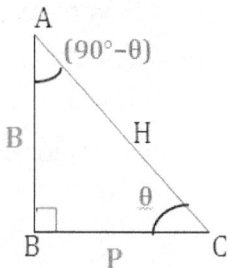

The relation between the sides H, P and B is:

$$H^2 = P^2 + B^2$$

$$or \ H = \sqrt{P^2 = B^2}$$

9.2 Trigonometric Functions

The ratios of the sides of a right triangle are known as trigonometric ratios or trigonometric functions. Three common trigonometric ratios are:

- **sine** (denoted as **sin**),
- **cosine** (denoted as **cos**), and
- **tangent** (denoted as **tan**).

These trigonometric ratios are defined, for an angle θ, as:

$$\sin(\theta) = \sin\theta = \frac{P(\text{for }\theta)}{H(\text{for }\theta)} = \frac{B(\text{for }90° - \theta)}{H(\text{for }90° - \theta)} = \cos(90° - \theta)$$

$$\cos(\theta) = \cos\theta = \frac{B(\text{for }\theta)}{H(\text{for }\theta)} = \frac{P(\text{for }90° - \theta)}{H(\text{for }90° - \theta)} = \sin(90° - \theta)$$

$$\tan(\theta) = \tan\theta = \frac{P(\text{for }\theta)}{B(\text{for }\theta)} = \frac{B(\text{for }90° - \theta)}{P(\text{for }90° - \theta)} = \cot(90° - \theta)$$

Consider an example:

ΔABC is as depicted in the figure. Calculate sin θ, cos θ and tan θ.

$$\sin\theta = \frac{P}{H} = \frac{4}{5}$$

$$\cos\theta = \frac{B}{H} = \frac{3}{5}$$

$$\tan\theta = \frac{P}{B} = \frac{4}{3}$$

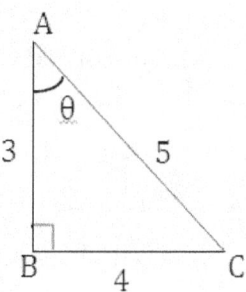

Note that tan $\theta = \dfrac{P}{H} = \dfrac{P/H}{B/H} = \dfrac{\sin\theta}{\cos\theta}$

There are three more trigonometric ratios defined by taking the reciprocal of the above defined trigonometric ratios:

- **cosecant** (denoted as **cosec**),
- **secant** (denoted as **sec**), and
- **cotangent** (denoted as **cot**).

These trigonometric ratios are defined, for an angle θ, as:

$$\text{cosec}(\theta) = \text{cosec }\theta = \frac{1}{\sin\theta} = \frac{H(\text{for }\theta)}{P(\text{for }\theta)} = \frac{H(\text{for }90° - \theta)}{B(\text{for }90° - \theta)} = \sec(90° - \theta)$$

$$\sec(\theta) = \sec\theta = \frac{1}{\cos\theta} = \frac{H\,(\text{for}\,\theta)}{B\,(\text{for}\,\theta)} = \frac{H\,(\text{for}\,90^\circ - \theta)}{P\,(\text{for}\,90^\circ - \theta)} = \cos ec\,(90^\circ - \theta)$$

$$\cot(\theta) = \cot\theta = \frac{1}{\tan\theta} = \frac{B\,(\text{for}\,\theta)}{P\,(\text{for}\,\theta)} = \frac{P\,(\text{for}\,90^\circ - \theta)}{B\,(\text{for}\,90^\circ - \theta)} = \tan\,(90^\circ - \theta)$$

Note that $\cot\theta = \dfrac{B}{P} = \dfrac{B/H}{P/H} = \dfrac{\cos\theta}{\sin\theta} = \dfrac{(1/\sin\theta)}{(1/\cos\theta)} = \dfrac{\cos ec\,\theta}{\sec\theta}$

Calculating the missing sides of triangles:

The missing sides of a right triangle can be calculated with the help of the trigonometric ratios.

For example: If Δ ABC is given as shown below. Find the side AC, given that sin θ = 0.7071 and BC = 4 cm.

BC = 4 cm

$$\sin\theta = \frac{P}{H} = \frac{BC}{AC} = \frac{4}{AC}$$

Thus, $AC = \dfrac{4}{\sin\theta} = \dfrac{4}{0.7071} = 5.66\,\text{cm}$

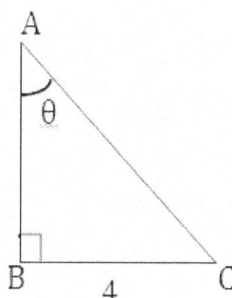

9.2.1 Trigonometric Ratios for certain angles:

At θ = 0°, B = H and P = 0 ($H^2 = P^2 + B^2$)

$$\sin\theta = \sin 0^\circ = \frac{P}{H} = \frac{0}{H} = 0$$

$$\cos\theta = \cos 0^\circ = \frac{B}{H} = 1 \qquad (\text{since } B = H)$$

$$\tan\theta = \tan 0^\circ = \frac{P}{B} = \frac{0}{B} = 0$$

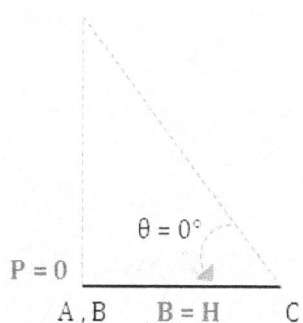

$$\cos ec\,\theta = \cos ec\,0^\circ = \frac{H}{P} = \frac{H}{0} = \infty$$

$$\sec\theta = \sec 0^\circ = \frac{H}{B} = 1 \qquad (\text{since } B = H)$$

$$\cot\theta = \cot 0^\circ = \frac{B}{P} = \frac{B}{0} = \infty$$

At θ = 90°, P = H and B = 0 ($H^2 = P^2 + B^2$)

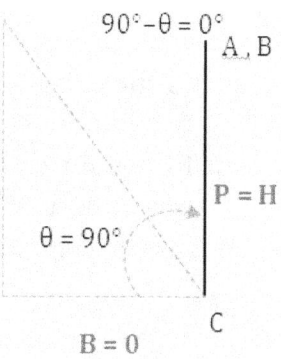

$$\sin\theta = \sin 90^\circ = \frac{P}{H} = 1 \qquad (\text{since } P = H)$$

$$\cos\theta = \cos 90^\circ = \frac{B}{H} = \frac{0}{H} = 0$$

$$\tan \theta = \tan 90° = \frac{P}{B} = \frac{P}{0} = \infty$$

$$\operatorname{cosec} \theta = \operatorname{cosec} 90° = \frac{H}{B} = 1 \quad \text{(since P = H)}$$

$$\sec \theta = \sec 90° = \frac{H}{B} = \frac{H}{0} = \infty$$

$$\cot \theta = \cot 90° = \frac{B}{P} = \frac{0}{P} = 0$$

θ	0°	30°	45°	60°	90°
$\sin \theta$	0	$\frac{1}{2}$	$\frac{1}{\sqrt{2}}$	$\frac{\sqrt{3}}{2}$	1
$\cos \theta$	1	$\frac{\sqrt{3}}{2}$	$\frac{1}{\sqrt{2}}$	$\frac{1}{2}$	0
$\tan \theta$	0	$\frac{1}{\sqrt{3}}$	1	$\sqrt{3}$	∞
$\operatorname{cosec} \theta$	∞	2	$\sqrt{2}$	$\frac{2}{\sqrt{3}}$	1
$\sec \theta$	1	$\frac{2}{\sqrt{3}}$	$\sqrt{2}$	2	∞
$\cot \theta$	∞	$\sqrt{3}$	1	$\frac{1}{\sqrt{3}}$	0

9.2.2 Signs of Trigonometric Ratios in IV Quadrants:

The signs of the trigonometric ratios in the IV quadrants are depicted by the following figure.

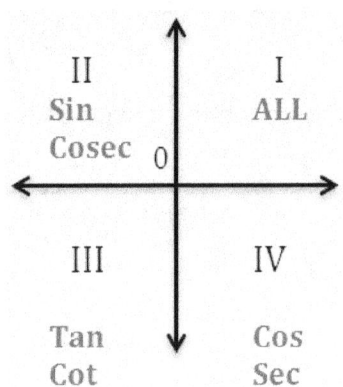

- In the first quadrant (I),
  all trigonometric ratios are positive
- In the second quadrant (II),
  sin and its reciprocal, cosec are positive
- In the third quadrant (III),
  tan and its reciprocal, cot are positive
- In the fourth quadrant (IV),
  cos and its reciprocal, sec are positive

It can be remembered as "CAST"

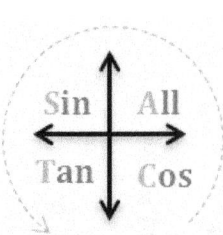

9.2.3 Trigonometric Formulas:

$-1 \leq \sin \theta \leq 1$

$-1 \leq \cos \theta \leq 1$

$-\infty \leq \tan \theta \leq \infty$

$\operatorname{cosec} \theta \leq -1$ or $\operatorname{cosec} \theta \geq 1$

$\sec \theta \leq -1$ or $\sec \theta \geq 1$

$-\infty \leq \cot \theta \leq \infty$

$\sin^2 \theta + \cos^2 \theta = 1$

$\tan^2 \theta + 1 = \sec^2 \theta$

$\operatorname{co}^2 \theta + 1 = \operatorname{cosec}^2 \theta$

$\sin (-\theta) \quad = -\sin(\theta)$

$\cos (-\theta) \quad = +\cos(\theta)$

$\tan (-\theta) \quad = -\tan(\theta)$

$\operatorname{cosec} (-\theta) = -\operatorname{cosec}(\theta)$

$\sec (-\theta) \quad = +\sec (\theta)$

$\cot (-\theta) \quad = -\cot (\theta)$

$\sin (90° - \theta) \quad = \cos(\theta)$

$\cos (90° - \theta) \quad = \sin(\theta)$

$\tan (90° - \theta) \quad - \cot(\theta)$

$\cot (90° - \theta) \quad = \tan(\theta)$

$\sec (90° - \theta) \quad = \operatorname{cosec}(\theta)$

$\operatorname{cosec} (90° - \theta) = \sec(\theta)$

$\sin (360° + \theta) \quad = \sin(\theta)$

$\cos (360° + \theta) \quad = \cos(\theta)$

$\tan (360° + \theta) \quad = \tan(\theta)$

$\cot (360° + \theta) \quad = \cot(\theta)$

$\sec (360° + \theta) \quad = \sec(\theta)$

$\operatorname{cosec} (360° + \theta) = \operatorname{cosec}(\theta)$

$\sin(2\theta) = 2\sin(\theta)\cos(\theta)$

$$= \frac{2\tan\theta}{1+\tan^2\theta}$$

$\cos(2\theta) = \cos^2(\theta) - \sin^2(\theta)$

$$= 1 - 2\sin^2(\theta)$$

$$= 2\cos^2(\theta) - 1$$

$$= \frac{2-\tan^2\theta}{1+\tan^2\theta}$$

$$\tan(2\theta) = \frac{2\tan\theta}{1-\tan^2\theta}$$

$$\sin = \pm\sqrt{\frac{1-\cos\theta}{2}}$$

$$\cos = \pm\sqrt{\frac{1+\cos\theta}{2}}$$

$$\tan = \pm\sqrt{\frac{1+\cos\theta}{1-\cos\theta}} = \frac{\text{Sin}\theta}{1+\cos\theta} = \frac{1-\cos\theta}{\sin\theta} = \cosec\,\theta - \cot\theta$$

$$\cot = \pm\sqrt{\frac{1+\cos\theta}{1-\cos\theta}} = \frac{\text{Sin}\theta}{1-\cos\theta} = \frac{1+\cos\theta}{\sin\theta} = \cosec\,\theta + \cot\theta$$

$\sin(A+B) = \sin A \cos B + \cos A \sin B$

$\sin(A-B) = \sin A \cos B - \cos A \sin B$

$\cos(A+B) = \cos A \cos B - \sin A \sin B$

$\cos(A-B) = \cos A \cos B + \sin A \sin B$

$$\tan(A+B) = \frac{\tan A + \tan B}{1-\tan A \tan B}$$

$$\tan(A-B) = \frac{\tan A - \tan B}{1+\tan A \tan B}$$

$$\cot(A+B) = \frac{\cot A \cot B - 1}{\cot B + \cot A}$$

$$\cot(A-B) = \frac{\cot A \cot B + 1}{\cot B - \cot A}$$

$$\sin A + \sin B = 2\sin\frac{(A+B)}{2}\cos\frac{(A-B)}{2}$$

$$\sin A - \sin B = 2\cos\frac{(A+B)}{2}\sin\frac{(A-B)}{2}$$

$$\cos A + \cos B = 2 \cos \frac{(A+B)}{2} \cos \frac{(A-B)}{2}$$

$$\cos A - \cos B = -2 \sin \frac{(A+B)}{2} \sin \frac{(A-B)}{2}$$

9.3 Inverse Trigonometric Functions

Corresponding to the trigonometric functions discussed in the previous section, there exist inverse trigonometric functions:

- **sine⁻¹** (denoted as **sin⁻¹**)
- **cosine⁻¹** (denoted as **cos⁻¹**)
- **tangent⁻¹** (denoted as **tan⁻¹**)
- **cosecant⁻¹** (denoted as **cosec⁻¹**),
- **secant⁻¹** (denoted as **sec⁻¹**), and
- **cotangent⁻¹** (denoted as **cot⁻¹**) .

These inverse trigonometric ratios are defined as:

$$\sin \theta = \frac{P}{H} \rightarrow \theta = \sin^{-1} \frac{P}{H}$$

$$\cos \theta = \frac{B}{H} \rightarrow \theta = \cos^{-1} \frac{B}{H}$$

$$\tan \theta = \frac{P}{B} \rightarrow \theta = \tan^{-1} \frac{P}{B}$$

$$\mathrm{cosec}\, \theta = \frac{H}{P} \rightarrow \theta = \mathrm{cosec}^{-1} \frac{H}{P}$$

$$\sec \theta = \frac{H}{B} \rightarrow \theta = \sec^{-1} \frac{H}{B}$$

$$\cot \theta = \frac{B}{P} \rightarrow \theta = \cot^{-1} \frac{B}{P}$$

Note: $\sin^{-1}(x)$ is not the same as $\dfrac{1}{\sin x}$, i.e. **−1** is not an exponent.

Calculating the missing angles of triangles:

The missing angles of a right triangle can be calculated with the help of the inverse trigonometric ratios.

For example: If Δ ABC is given as shown below. Find the angle ∠A, given that BC = 4 cm and AB = 3 cm.

BC = 4 cm

AB = 3 cm

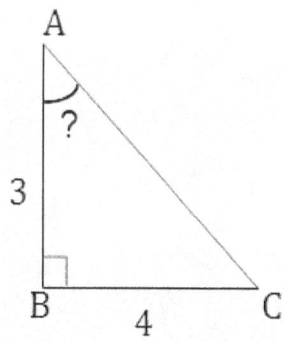

$$\tan \angle A = \frac{P}{B} = \frac{BC}{AB} = \frac{4}{3}$$

$$\angle A = \tan^{-1}\left(\frac{4}{3}\right)^{\circ}$$

9.3.1 Inverse Trigonometric Formulas:

$\sin^{-1}(\sin \theta) \qquad = \theta$

$\cos^{-1}(\cos \theta) \qquad = \theta$

$\tan^{-1}(\tan \theta) \qquad = \theta$

$\operatorname{cosec}^{-1}(\operatorname{cosec} \theta) = \theta$

$\sec^{-1}(\sec \theta) \qquad = \theta$

$\cot^{-1}(\cot \theta) \qquad = \theta$

$\sin(\sin^{-1} x) \qquad = x$

$\cos(\cos^{-1} x) \qquad = x$

$\tan(\tan^{-1} x) \qquad = x$

$\operatorname{cosec}(\operatorname{cosec}^{-1} x) = x$

$\sec(\sec^{-1} x) \qquad = x$

$\cot(\cot^{-1} x) \qquad = x$

$$\operatorname{cosec}^{-1} x = \sin^{-1} \frac{1}{x}$$

$$\sec^{-1} x = \cos^{-1} \frac{1}{x}$$

$$\cot^{-1} x = \tan^{-1} \frac{1}{x}$$

$$\sin^{-1} x + \cos^{-1} x = \frac{\pi}{2}$$

$$\tan^{-1} x + \cot^{-1} x = \frac{\pi}{2}$$

$$\operatorname{cosec}^{-1} x + \sec^{-1} x = \frac{\pi}{2}$$

$$\left.\begin{array}{l} \sin^{-1}(\cos\theta) \ = \dfrac{\pi}{2} - \theta \\[3mm] \cos^{-1}(\sin\theta) \ = \dfrac{\pi}{2} - \theta \\[3mm] \tan^{-1}(\cot\theta) \ = \dfrac{\pi}{2} - \theta \end{array}\right\} \ 0 \le \theta \le \pi$$

$$\cot^{-1}(\tan\theta) \ = \dfrac{\pi}{2} - \theta,\ 0 \le \theta \le \pi$$

$$\left.\begin{array}{l} \sec^{-1}(\operatorname{cosec}\theta) = \dfrac{\pi}{2} - \theta \\[4mm] \operatorname{cosec}^{-1}(\sec\theta) = \dfrac{\pi}{2} - \theta \end{array}\right\} \ 0 \le \theta \le \pi/2$$

$$\sin^{-1}(-x) \ = -\sin^{-1}x$$

$$\cos^{-1}(-x) \ = \pi - \cos^{-1}x$$

$$\tan^{-1}(-x) \ = -\tan^{-1}x$$

$$\operatorname{cosec}^{-1}(-x) = -\operatorname{cosec}^{-1}x$$

$$\sec^{-1}(-x) \ = \pi - \sec^{-1}x$$

$$\cot^{-1}(-x) \ = \pi - \cot^{-1}x$$

$$\sin(\cos^{-1}x) \ = \sqrt{1-x^2}$$

$$\cos(\sin^{-1}x) \ = \sqrt{1-x^2}$$

$$\sin(\tan^{-1}x) \ = \dfrac{x}{\sqrt{1+x^2}}$$

$$\cos(\tan^{-1}x) \ = \dfrac{x}{\sqrt{1+x^2}}$$

$$\tan(\sin^{-1}x) \ = \dfrac{x}{\sqrt{1-x^2}}$$

$$\tan(\cos^{-1}x) \ = \dfrac{\sqrt{1-x^2}}{x}$$

Exercises

1. Multiplication Tables

1.1 (a) $2 \times 4 = ?$ (b) $12 \times 8 = ?$ (c) $7 \times 9 = ?$ (d) $18 \times 6 = ?$

(e) $10 \times 8 = ?$ (f) $15 \times 5 = ?$ (g) $19 \times 4 = ?$ (h) $17 \times 9 = ?$

1.2 (a) ০ × ৪ = ?   (b) ৬ × ৫ = ?   (c) ৫ × ৫ = ?   (d) ৯৯ × ৬ = ?

(e) ১৯ × ৫ = ?   (f) ১৪ × ৬ = ?   (g) ৮ × ৮ = ?   (h) ১৩ × ৫ = ?

2. Number Theory

2.1 (a) Identify the natural number(s): 0, −10, 6, 0.14, 10000, $\dfrac{2}{3}$

(b) Identify the whole number(s): −15, 600, 2, 0, $\dfrac{6}{7}, \dfrac{16}{4}$

(c) Identify the integers(s): −15, 0, 0.2, −1.75, $\dfrac{6}{7}$, 100

(d) Identify the rational number(s): −1, 0, −6.45, $\dfrac{8}{7}$, $\sqrt{2}$

(e) Identify the irrational numbers(s): −100, $\sqrt{5}$, $\sqrt[3]{10}$, −10.76, $\dfrac{6}{7}$, π

2.2 Calculate the following operations:

(a) $16 + 3 = ?$ (b) $23 + 7 = ?$ (c) $17 - 7 = ?$ (d) $24 \times 5 = ?$

(e) $50 \div 5 = ?$ (f) $200 - 18 = ?$ (g) $66 \div 11 =$ (h) $10^2 = ?$

(i) $\sqrt{4} = ?$ (j) $5^3 = ?$ (k) $(2 + 4) \div 6 + 3 = ?$

(l) $2 + 6 \div 3 + 3 = ?$ (m) $500 \div 10^2$

2.3 Classify into odd and even numbers: −3 , 0 , 1, 4, 100, 256, 1799

2.4 Calculate the following operations:

(a) $-16 + 5 = ?$ (b) $-4 \times -5 = ?$ (c) $5 + -1 = ?$ (d) $12 \times -6 = ?$

(e) $-15 \div 5 = ?$ (f) $18 \div -10 = ?$ (g) $-100 \times 7 = ?$

2.5 Convert the following decimal numbers into fraction:

(a) 1.25   (b) −2.5   (c) 66.6   (d) 100.5   (e) −15.5

2.6 Identify the prime numbers: 1, 17, 2, 9, 11, 23, 77, 13, 125, 41

2.7 Calculate the prime factors of : (a) 23   (b) 55   (c) 24   (d) 105

2.8 Find the highest common factor (HCF) of:

    (a) 12, 15   (b) 20, 30   (c) 14, 12   (d) 8, 10, 3

2.9 Find the least common multiple (LCM) of:

    (a) 5, 7   (b) 6, 10   (c) 50, 70   (d) 3, 9, 21

2.10 Identify the pair of co-primes:

    (a) 14, 15   (b) 15, 9   (c) 15, 28

3. Roman Numerals

3.1 Convert the following from Hindu Arabic to Roman numeral system:

    (a) 1850   (b) 16   (c) 345   (d) 500

3.2 Convert the following from Roman numeral to Hindu Arabic Numeral System:

    (a) XIX   (b) XCVII   (c) XXVIII   (d) DCCLXXXI

4. Algebra

4.1 Solve the following equations for unknown/variable:

    (a) $4x - 3 = 5$

    (b) $x + 6 = 2x - 10$

    (c) $\dfrac{y}{3} + 5 = 10$

    (d) $2z - 3 = \dfrac{z}{3} + 3$

4.2 List the coefficients of x in the following equations:

    (a) $4.5x - 5 = 10$

    (b) $3.5 - 2x = 7$

4.3 Solve the following equations with two variables x and y:

    (a) $3x - 5 = 10$

        $x + y = 15$

    (b) $x + 2y = 10$

        $x + y = 6$

4.4 Identify whether the following are polynomials or not? If yes, calculate their degrees:

    (a) $2x^2 - 3x^4 + 5$

    (b) $5y^2 + 3y^{-2}$

    (c) $(a + b)^2$

    (d) 2

    (e) $7z^2 + \sqrt{y}$

5. Mensuration

5.1 Find the area of the following triangles:

    (a) acute triangle with a base of 15 cm, altitude of 4 cm.

    (b) right triangle with a base of 4 inches and a height of 3 inches.

    (c) An equilateral triangle with side 4 cm

5.2 The area of a triangular mat, with the base as 2 feet, is 12 square feet. Find its height.

5.3 Calculate the missing side of the right triangles below:

(a)

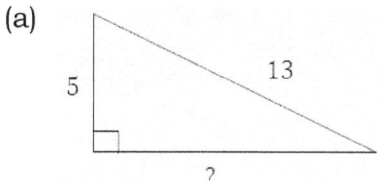

(b)

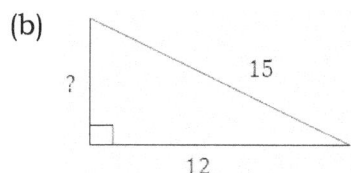

5.4 The length and breadth of a rectangular playground is 70 m and 35 m. Find the cost of leveling it at the rate of ₹ 2 per m². Also, find the distance covered in taking 6 rounds of the playground.

5.5 Find the circumference and area of a circle with diameter 6 cm.

5.6 The circumference of a circle exceeds its diameter by 30 cm. Find the radius of the circle. Take π = 22/7.

5.7 A cylinder has a base radius of 4 cm and a height of 10 cm. Calculate its volume, total surface area and curved surface area. Also calculate the surface area of the base of the cylinder. Take π = 3.14.

5.8 A toy is composed of a base that is a hemisphere with a conical top. The volume of the conical top is 30π cm³ and its height is 10 cm. Calculate the volume of the hemisphere, that is the base of the toy.

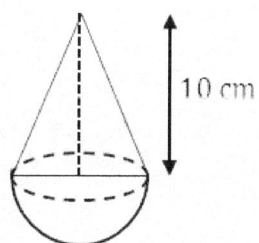

5.9 Calculate the perimeter, lateral surface area, total surface area and volume of a cube with side 3 ft.

5.10 The perimeter of an isosceles triangle is 16 cm. If one of the two equal sides is 6 cm, calculate the third side.

6. Conversion Units

6.1 Convert the following units:

(a) 6.5 hours to minutes

(b) 25 paisa into rupee

(c) 50 cm to km

(d) 46 inches to feet

(e) 3.5 radians to degree

(f) 60.56° in DMS form

6.2 Convert the following by analyzing the metric prefixes:

(a) 26 kg to Mg

(b) 45 kHz to DHz

(c) 2.7 dm to cm

(d) 1080 µg to mg

7. Trigonometry

7.1 Consider the Δ ABC as shown below. Calculate the side AB.

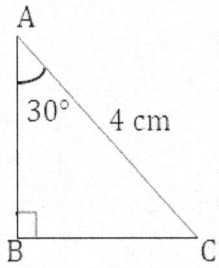

7.2 Consider the Δ ABC as shown below. Calculate the side AC, given that cos θ = 0.5.

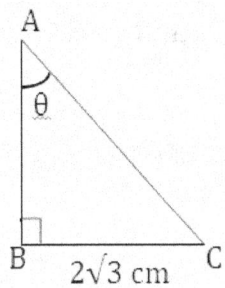

7.3 If Δ PQR is given as shown below. Calculate the angle ∠R.

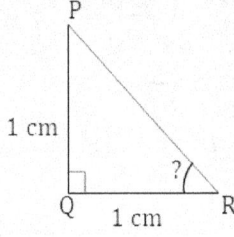

7.4 Calculate the following:

(a) sin (−30°)

(b) cos 120°

(c) tan 225°

(d) cosec 300°

(e) cot 135°

(f) sec (− 60°)

(g) tan (150°)

(h) sec 180°

7.5 Consider the Δ ABC as depicted below. Calculate sin θ, cos θ, tan θ, cosec θ, sec θ and cot θ .

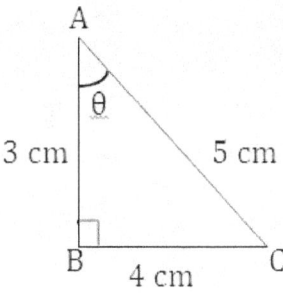

Answers:

1.1 (a) 8   (b) 96   (c) 63   (d) 108   (e) 80   (f) 75   (g) 76   (h) 153

1.2 (a) ০   (b) ৩৫   (c) ২৫   (d) ৬৬   (e) ৯৬৯   (f) ২৮   (g) ৬৫   (h) ৯৯৬

2.1 (a) 6 and 10000   (b) 600, 2, 0 and $\frac{16}{4}$ = 4 (c) −15, 0 and 100

   (d)−1, 0, −6.45 and $\frac{8}{7}$ (e) √5, $\sqrt[3]{10}$ and π

2.2 (a) 19   (b) 30   (c) 10   (d) 120   (e) 10, with remainder 0   (f) 182
   (g) 6,with remainder 0   (h) 100   (i) 2   (j) 125   (k) 4   (l) 7
   (m) 5, with remainder 0

2.3 Odd: −3, 1, 1799 and Even: 0, 4, 100, 256

2.4 (a) −11   (b) 20   (c) 4   (d) −72   (e) −3   (f) −1, with remainder 8   (g) −700

2.5 (a) $\frac{5}{4}$ (b) $-\frac{5}{2}$ (c) $\frac{333}{5}$ (d) $\frac{201}{2}$ (e) $-\frac{31}{2}$

2.6 17, 2, 11, 23, 13, and 41 are prime numbers

2.7 (a) 23   (b) 5 and 11   (c) 2, 2, 2 and 3   (d) 3, 5 and 7

2.8 (a) 3   (b) 10   (c) 2   (d) 1

2.9 (a) 35   (b) 30   (c) 350   (d) 63

2.10 (a) and (c) are co-prime pairs; (b) is not.

3.1 (a) MDCCCL   (b) XVI   (c) CCCXLV   (d) D

3.2 (a) 19   (b) 97   (c) 28   (d) 781

4.1 (a) x = 2   (b) x = 16   (c) y = 15   (d) z = 4

4.2 (a) 4.5   (b) 2

4.3 (a) x = 5 and y = 10   (b) x = 2 and y = 4

4.4 (a) Polynomial, degree 4   (b) Not polynomial   (c) Polynomial, degree 2   (d) Polynomial, degree 0   (e) Not polynomial

5.1 (a) 30 cm²   (b) 6 inch²   (c) 4√3 cm²

5.2 Height is 12 feet

5.3 (a) 12   (b) 9

5.4 Distance covered in taking 6 rounds of the playground = 1260 m Cost of leveling = ₹4900

5.5 Circumference = 18.84 cm and Area = 28.26 cm²

5.6 Radius = 7 cm

5.7 Volume = 502.4 cm³; Total Surface Area = 351.68 cm²; Curved Surface Area = 251.2 cm²; Surface area of base = 50.24 cm²

5.8 Volume of the hemisphere = 18π cm³ = 56.52 cm³

5.9 Perimeter = 36 ft; Lateral Surface Area = 36 ft²; Total Surface Area = 54 ft²; Volume = 27 ft³

5.10 The third side = 4 cm

6.1 (a) 390 minutes   (b) Rs. 0.25   (c) 0.0005 km   (d) 3.833 feet   (e) 200.53°   (f) 60° 33' 36"

6.2 (a) 0.026 Mg   (b) 4500 DHz   (c) 27 cm   (d) 1.08 mg

7.1 AB = 2√3 cm

7.2 AC = 4 cm

7.3 $\angle R = \tan^{-1} (1) = 45°$

7.4 (a) $-\dfrac{1}{2}$ (b) $-\dfrac{1}{2}$ (c) 1   (d) $-\dfrac{2}{\sqrt{3}}$ (e) -1 (f) 2   (g) $-\dfrac{1}{\sqrt{3}}$ (h) -1

7.5 $\sin \theta = \dfrac{4}{5}$; $\cos \theta = \dfrac{3}{5}$; $\tan \theta = \dfrac{4}{3}$; $\operatorname{cosec} \theta = \dfrac{5}{4}$; $\sec \theta = \dfrac{5}{3}$; $\cot \theta = \dfrac{3}{4}$

www.ingramcontent.com/pod-product-compliance
Lightning Source LLC
Chambersburg PA
CBHW081728220526
45468CB00008B/2015